The Sociology of Invention

By the Same Author

Supplement, 1970, to The Sociology of Invention; to be published.

Inventing the Ship. The original companion volume of **The Sociology of Invention**, exemplifying its 38 principles thru consideration of all the more important inventions in the history of the merchant ship. xvi +294 pp., 80 ills., $5.00, available from The M.I.T. Press.

Social Implications of Technical Advance. Vol. I, No. 4 of **Current Sociology** (UNESCO, 1953). 81 pp., out of print.

Invention and the Patent System (Government Printing Office, 1964). Contains much material supplemental to the present volume. 247 pp., $0.60.

An Ugly Duckling's Swan Song. The author's autobiography, to be published serially in **Sociological Abstracts** and later as part of a book collection of autobiographies of sociologists.

The Sociology of Invention

An essay in the social causes, ways and effects of technic invention, especially as demonstrated historicly in the author's **Inventing the Ship.**

S. C. GILFILLAN

(Columbia) Sometime Instructor of social sciences, in the colleges of Grinnell and Sewanee, and Curator of Ships, and of Social Sciences, in the Museum of Science and Industry, Chicago.

THE M.I.T. PRESS

CAMBRIDGE, MASSACHUSETTS AND LONDON, ENGLAND

FIRST M.I.T. PRESS PAPERBACK EDITION MARCH 1970

SBN 262 07009 X (HARDCOVER)
SBN 262 57020 3 (PAPERBACK)

LIBRARY OF CONGRESS CATALOGUE CARD NUMBER: 75-97497

TO MY OWN FRIENDS, FELLOW STUDENTS OF THE SOCIAL ASPECTS OF INVENTION

L. J. Carr	Wm. F. Ogburn
Z. C. Dickinson	Joseph Rossman
Waldemar Kaempffert	B. S. Sanders
R. K. Merton	C. P. Wright

PREFACE

That *invention* is an important subject for modern mankind to understand and perhaps later improve, all will agree. That invention is partly caused, hampered, promoted, steered by *social* factors and institutions (such as wealth, or the patent system), and not simply by developments in the physical sciences and industrial practice, will also be allowed. (How great is the social causation we shall discuss.) Likewise that inventions have wide social, and not simply industrial *effects,* has been common knowledge for nigh a century. There would seem then every call for a treatise on the Sociology of Invention. Yet not one book with this definite field has been published in any language, unless one could so describe Dr. Rossman's excellent work on *The Psychology of the Inventor.*[171] Many useful articles and parts of books have been written, and are listed in our Bibliographies and Notes. But a formal study like the present one, especially on the side of social causation, pulling the subject together and proposing a list of definite social principles of invention, seems all the more needed, the more the way has been prepared for it by intelligent but inconclusive, discordant and scattered writings.

The social aspects of invention have commonly been treated by two very different classes of men (with a few honorable exceptions (ftN*1*), utterly different in the thot worlds from which they start, and each needing the knowledge of the other for sound conclusions. One group are the inventors, engineers, physical scientists and pat-

1 Let us cite notably Dr Rossman, acquainted with Psychology, social science, Chemistry, invention and patents; Vaughan (Note 180 in our Appendix), Ogburn (N 169 & 237), Epstein (N 193), Usher (N 179) and Ravenshear (N 170).

ent lawyers, knowing much more than the present writer about the craft of invention, its sciences, industries, ways, traditions and the practical problems that the inventor recognizes. But this group possess no social science, they do not even know when they stray into its garden, and their facile pronouncements on the social effects and causes of invention are normally traditional and without value. The other group are the social scientists—Sociologists, Economists, Historians etc. They similarly stray into Engineering history without realizing that they are in danger of wading beyond their depth, and they are likewise free with generalizations about inventions, which are untested, unreliable, and being commonly based on popular engineering stories of famous inventors and their deeds in by-gone days, lack verity to the average inventive realities of today, these social scientists being little informed as to the physical sciences and the business of modern invention. Two other classes of writers on invention might be mentioned and dismist—the business leaders and the popular magazine writers, who knowing neither the engineering nor the social sciences, can only make themselves entertaining, with traditional laudations of the might and glory of invention.

Our problem is to combine those two worlds of thot, which have so rarely been conjoined—social science and engineering—in order to produce a rather new and fertile hybrid, a Sociology of Invention. It is a difficult problem, this getting people to study and think in an unfamiliar world; and we have tried to solve it for some readers, and dodge it for most, by a new plan:

For factual evidence and inductive basis, in the initial facts of engineering, it is the rule for all writers on invention to cull offhand the accounts of historic inventions, as we said, and present one, two or a few instances as proof of each generalization claimed. These examples are chosen for their clear, pleasing and striking nature, hence usually are extreme instances of the rule claimed. How many ignored exceptions protest the law propounded, the reader can know only by deep learning and thot. Yet this method may serve, when the readers are familiar with the facts, and when the writer is highly so and a trustworthy scientist. For good or ill there is plenty of it in our own volume. But for sturdy proof and scientific advance a method is needed which has completeness, considering the exceptions along with the conformities. Statistical studies do this, and such have only recently been made, on some aspects of invention, by Rossman, Carr, Hart, Dickinson, Hulme and others listed in our appended Bibliographies on Social Causation, and Results. We shall add a few tables, graphs and informal counts and averages ourselves. But statistics can explain nothing by themselves, and are hard to use aright with this subject. Patent statistics are questionable measures of minor invention, and not available to count the major inventions which most interest us.

The *case study* method is one which can be used with a certain completeness, for sound results on the social process of invention, and it has hardly been tried before. It is one-sided, limited to one chosen field of invention. But that may be a real advance over the complete, statistical method, because it permits the writer and even the reader to *understand* the field being talked about, whereas no one, least of all a social scientist, can understand all fields, all inventions. If one choose, arbitrarily but not unreasonably, a certain category of inventions, as we have in the ship, and then try to take up and understand *all* the inventions and facts falling within that category, one should be able to demonstrate some valid inductions, based on a representative section of the *whole* truth, showing the discrepancies from the proposed laws as well as the conformities. That would be much better than to rely on a few instances, selected to prove something.

So we have chosen for our case study and inductive material, the history of the merchant ship, between floating log and rotorship, as to all the important inventions made in the lines of descent so delimited, and a few inventions straying around the edges. The ship was chosen for these reasons: She is sufficiently familiar to the writer. She has an extremely ancient as well as uninterrupted and modern history. There is a comparative abundance of historical material on her. She takes a good picture. She is beloved by millions. She involves little chemistry or electricity, and usually only mechanics, understandable to the lay reader, if he will but think. She is probably a fairly good examplar of the usual social principles of invention, altho probably more conservative and evolutionary than the average, from the separation and danger of her often lonely position on the storm-swept sea. Finally, ships have been of great importance, especially to the progress of civilization, thru linking the ideas and peoples of distant countries, and supplying cheap transportation, a fundamental necessity of centralized production, specialization and progress.

We have placed our case study of the ship in a separate work, called *Inventing the Ship,* in order that students of invention who feel seasick when they even hear of a ship may not have to buy or look at our evidence. But we have made it all understandable to the intelligent layman, with a little thot, and occasionally a dictionary. It was published in shorter form in the *Marine News.* And we have well indexed it, filled it with pictures and sought to make it interesting. We hope, therefore, that many of our present readers will examine the evidence in our companion volume, to see if it bears out our generalizations, and perhaps to find other social laws of invention for themselves, and to obtain instances and further discussion on many points. The other readers, who have not time, or stomach, to look into a technic history to test our evidence, will kindly take on faith that the history of the ship bears out our claims.

A little about the author would be appropriate in any book, to advise the reader of his fields of greater and less competence, and of his biases. The present writer cannot qualify as an inventor, engineer nor patent lawyer, despite some efforts in the first two fields; and he is very little of a mariner. He is a sociologist, interested especially in the history and prophecy of the developments of civilization, and in the social aspects of invention, on which he has done further work; but he claims also some knowledge of nautical and mechanical matters. A Bibliography of his publications is in an Appendix.

He has also obtained appreciated help on various aspects of this volume from the following authorities, aside from their publications: Joseph *Rossman,* Ph.D., M.P.L., LL.B., B.S. in Chem. Eng., patent examiner and Editor of the *Journal of the Patent Office Soc.* (cf. 2nd par. below); Prof. Lowell J. *Carr,* who read the first part of the book; Profs. Alvan A. *Tenney,* John W. *Oliver,* and Wm. F. *Ogburn,* who read the whole, and the last of whom gave instruction and employment on the social effects of invention; Waldemar *Kaempffert,* Esq., for illumination and employment anent invention; Barkev S. *Sanders,* Ph.D., for many suggestions on invention, and assistance with statistics; Prof. Z. C. *Dickinson,* for general instruction; Ellsworth *Huntington,* Ph.D., Samuel P. *Hunt,* M.A., C.E., L. J. Lesh, inventor, and Bernhard J. *Stern,* Ph.D., for criticism; and Mrs. Louise Wenzel *Gilfillan,* A.M., for greatest assistance and encouragement.

Two kinds of Notes, in a single numerical series, are used in this as in our other volume. Those which continue a discussion are placed at the foot of the page, and referred to by exponential italic numerals, thus,[12] or in the Index, etc. as ftN*12*. Less interesting notes, usually confined to citations, are assembled in the first Appendix, for readier cross reference, and are indicated in the text by small figures of the exponential type, thus [13], and in the Index etc. as N 13, using ordinary, not italic numerals. The series is continued with such numerals in the Bibliographies.

This book, except for certain additions and corrections, was published serially in the *Journal of the Patent Office Society,* with the title Social Principles of Invention. My great thanks are due to its learned and progressive editor, Dr. Joseph Rossman, for this dissemination of it in the important circle of those concerned with patents, and for permission to use the same type for this book, and for general counsel in this field where he has done so much. *Journal* readers now using this book will find the numbers of the Principles all changed about and subdivided and grouped, to obtain a logical arrangement, according to the following table:

Old	New	Old	New	Old	New	Old	New	Old	New
1	1	9	23b	17	18	24	14	32	30
2	26	10	34	18	19	25	24	33	10
3	27	11	37	19	20	26	36	34	22
4	2	12	17	19B	21	27	2	35	8
5	6	13	29	20	32	28	3	36	38
6	7	14	33	21	35	29	11	37	6
7	25	15	15	22	13	30	12	38	9
8	23a,28	16	16	23	31	31	5	39	4

5623 Blackstone ave.,
Chicago; May, 1935.

TABLE OF CONTENTS

THE SOCIOLOGY OF INVENTION

CHAPTER I

A PREVIEW OF 38 SOCIAL PRINCIPLES OF INVENTION

Generalizations are supposed to follow after the facts on which they are based; and so may most of these thirty-eight, because our companion volume *Inventing the Ship* contains evidence supporting at least 30 of them. For the rest, Numbers 10, 17-21, 28, 37 and 38, evidence from marine history was not to be had, but they will be supported hereinafter, from other fields of man's progress. At the same time the other Principles will be critically discust and demonstrated, in a convenient order for taking up, differing from the numerical order adopted below as most suitable for a logical conspectus.

But first a little general discussion of our *1st Principle* seems especially needed, because the very essence of invention is commonly misunderstood, not alone by the laity, but often by inventors, engineers and sociologists. Above all its outstanding characteristic, its evolutionary nature, its being almost wholly an age-old, multitudinous accretion of little details, modifications, perfectings, minute additions, is not fully appreciated by even the best of authorities.[2] As to inventions we are still in the antediluvian geologic age, holding a cataclysmic rather than evolutionary theory of the origin of things. The great inventions are supposed to have been made by certain great men, much as Adam was made and then leaned against a fence to dry, according to the song. Most of

[2] The first man to see this evolutionary, social, non-human character of invention, and the first social scientist to treat the subject, was perhaps the historian (V. E.) Philarète Chasles, in his *Etudes sur les premiers temps du Christianisme et sur le Moyen Age,* Paris 1847. The first social science book on invention was Yves Guyot: *L'Inventeur,* 1867, (N 156).

those creative heroes have been forgot, but we do remember the names of some (who seem to be chiefly Americans). They were taught us at our most receptive age, by grammar school teachers at the behest of educational authorities who knew well how to inculcate American patriotism. So here, for instance, is Rear-Admiral Bradley A. Fiske, LL.D., truly a distinguished naval officer and inventor, author of *Invention, the Master-key to Progress* (1921), saying in it that to Daedalus are ascribed the saw, gimlet, plumb-line, ax, wedge, lever, masts and sails.[3] "As no record shows to us that the inventions just enumerated (except masts and sails) had been invented elsewhere, we may feel justified in inferring that they were invented in Greece by Dædalus, or by some other man bearing a different name—or by some other men." As a matter of fact, not one of those was invented in Greece, but all previously, and never by single inventors anywhere.[4] In some of his other passages, to be sure, Admiral Fiske reflects a sound view.

The ship is an especially good illustration of technic evolution; and indeed the more scientific of the nautical authorities have rather taken this trait for granted. Ship designers have been particularly averse to revolutionary innovations, because of the danger to lives and much property, that is likely to follow any mistake sent to sea, or set afloat on any waters. However, this peculiarity of ships has probably been weakened in the modern age, with its growth of safety, its stirring together of intelligences, and subjecting to standard customs of science all directive and engineering men of whatever industry. And to this modern age we give most of our attention. Furthermore, other elaborate studies that follow down one line of invention, as of the saw[5], the steam engine, the house, clothes etc., show like evolutionary growth.

---

[3] B. A. Fiske: *Invention, the Master-key to Progress.* Dutton pub., N. Y. 1921; 365 pp.; p. 57. A valuable book, in spite of this naive view, which is well contradicted on p. 217, and Dædalus called probably a myth.

Many other Principles might be found, and some are more or less explicitly stated in our two volumes; but the following 38 seem those which can be best defended, as of universal modern application, at least as tendencies. A tendency, contrasting with an absolute law, does not require a visible conformity in every instance of its applicability. For instance, the law that water tends to flow down hill is universally true; and yet water often flows upward, because of various forces other than direct gravity . . . . Against each of these our Principles, the modern unaccountable exceptions seem few.

It is to be expected that the applicability of any social principles will change with changing civilization. Ours are propounded with respect to modern conditions, and so altho their terms give them a rather universal validity, we must avow that the applicability to pre-modern and especially to primitive conditions is questionable, of Principles 9, 11, 12, 23-5, 30, 32, 33 and 38b.

The pages wherein each Principle is explicitly mentioned may be ascertained from the Index, which distinguishes the chief discussion.

Social Principles of Invention

A. The Nature of Invention

1. (a) What is called an important invention is a perpetual **accretion** of little details, (b) probably having neither beginning, completion nor definable limits, (c) tho it is hazily and somewhat arbitrarily defined by a word or phrase in the English language, and by our standardizing habits in thot and industry, as well as in language. (d) An invention is an **evolution**, rather than a series of creations, and much resembles a **biologic** process, because it has a basic kinship with this, thru innate human mentality.

2. (a) An invention is essentially a **complex** of most diverse elements—a design for physical object, a process of working with it, the needed elements of science, if any; the constituent materials, a method for building it, the raw materials used in working it, such as fuel, accumulated capital such as factories and docks, with which it must be used, its crew with their skills, ideas and shortcomings, its financial backing and management, its purpose and use in conjunction with other sides of civilization and its popular evaluation. Most of these parts in turn have their separately variable elements. (b) A change in any one of the elements of the complex will alter, stimulate, depress or quite inhibit the whole.

3. (a) Invention is a **new combination** from the "prior art," i.e. ideas previously known from all the above categories, (b) having varying number, ripeness and recency.

4. Invention need not be based on **prior science.** It often precedes and evokes the apposit science.

5. Invention would become much easier with advancing civilization, were "invention" not defined by language usage, which sets the modern standard higher. Hence many "important inventions" of early times would be too simple and easy to be called inventions, if made today.

B. Changes Evoking Invention

6. (a) The achievements of the inventor himself, and of his compeers, are a major cause of changes in the milieu out of which is compounded the inventional complex described in Prin. 2. "Invention is (one) mother of necessity." (b) Some other chief **changes causing invention** are growths of wealth, education, population, industrialism, and commercial organization.

7. (a) The more **recent changes** in this milieu, commonly evidenced by a notable change in the price of something, or being a scientific discovery, or falling to hand of an adaptable invention, have two special significances; (1) Tho weak they may serve as "triggers;" and (2) they attract attention, stimulating inventive adjustment. (b) One or more recent changes may be identified as the trigger or the active causes of almost any invention.

C. The Rate of Growth and Life Cycle of an Invention

8. Inventions normally progress but slowly in the period before they become practical propositions.

9. The rate of **increase** in the **use** of an invention probably tends to decline continuously from the start of practical success, and with time enough will become a growing rate of decrease in use.

10. (a) An invention which saves anything, tends to **reduce its own importance** and insofar discourages further invention in the same line. (b) If, as usually, it is one out of a chain of steps toward a final production, it encourages that production and stimulates invention for each other step toward it. (c) Whether an economical invention will ultimately diminish or increase its own importance depends upon the size of its share in producing the final consumption good, and on the flexibility of demand for that good. (d)

An invention which saves nothing, but gives a new or improved product, is bound to increase its own field.

11. **Inventions persist**—(a) in the modern world they rarely pass totally out of use, still less out of memory. (b) Hence technology is cumulative, and the idea-fund for further new combinations continually increases.

12. But inventions are **rarely rejuvenated.** Once the progress of one has ceased it is rarely resumed in like nature.

13. The perfecting of a type of object mechanically is evidenced by its attainment of **beauty.**

14. Inventions having each their own trend, and their understandable causation, they can be and have been successfully **predicted** centuries ahead.

D. Factors Fostering, Retarding and Locating Invention

15. **Use promotes improvement.** The rate of progress on an invention tends to vary with the absolute rate at which that device is being turned out in practice. But there are so many other factors affecting the invention's progress that the action of this Principle is often negated.

16. Hence invention in each line centers in a few regions which most abundantly produce that line, these being responsible for about all the progress in it, and rest looking to the **fountain lands**, and rarely doing more than borrow and adapt their inventions to local needs.

17. Invention is helped by all **specializations** of labor (Spencerian integration resulting in a piling up of one kind of work at one point), whether between (a) regions, (b) firms or (c) workmen, since these all enable a more intensive, full use of the capital required to devise, build and operate an invention, especially one that would increase output. Further, (a) and (b) encourage a very helpful fourth form (d) of specialization, the coming of the professional inventor.

18. Leading firms. There is a tendency from the last three Principles for the firms upon whom the largest amount of a certain production has become centralized, to make the most inventions thereon, since they can exploit the invention most intensively and with least obstruction from the patent system, and they most need such inventions as are adaptations to a greater scale of production than heretofore reached.

19. But the **standardization** which tends to accompany wide organization obstructs inventions which would require changing the standard form.

20. (a) And **growth of capital** and use of **durable** materials contain one anti-inventive principle, since most novelties require the discarding of equipment, both that directly redesigned, and that which must interact with it, like trains and tracks. (b) Acquired specialized knowledge, workmen's skill and commercial good-will are durable forms of capital that similarly oppose invention.

21. A **rapid expansion** of population and industry, as in America hitherto, is **favorable** to invention, since it facilitates adding new equipment without scrapping old.

22. Increase of **population** and/or industry also stimulate invention, in that they increase the absolute need for an invention, and the number of finders, while the cost of finding it remains the same.

E. Principles of Chance

23. (a) The inventors are led by perceptions of the possibility and need of making easily enough some recombinations of elements of the "prior art" and milieu. (b) Invention by **accident** is very rare or unknown in marine history, and usually requires that the need for the invention be already felt by an inventive observer of the accident.

24. **Inevitability.** With the progress of the craft of invention, apparently a device can no longer remain unfound when the time for it is ripe, i.e. when the needed factors indicated in 2, 6 and 7 are present.

25. Hence simultaneous **duplicate invention** often occurs, and constantly would occur if not forestalled by news that the invention is made.

F. Inventors and other Classes, and Tendencies in the Craft

26. There is no indication that any individual's genius has been necessary to any invention that has had any importance. To the historian and social scientist the progress of invention appears **impersonal.**

27. Yet invention can only come at the hand of some sort of **inventors**, and its directions, frequency and efficiency are determined wholly thru deliberate actions by these men, in some proportion to their absolute numbers, intelligence, moral traits, strength of motives for inventing, time free for it, and mental and mechanical equipment for it.

28. Perception of the need and the way to meet it depends first upon any individuals of a numerous inventive class, but **indirectly** upon the suggestions tost up by a wider intelligent and technologicly minded class, more indirectly upon the whole population's thoughts, and **ultimately** upon the physical environment and the general social and racial heritage, which determine the mental level and slant of each class and country.

29. The inventors are checked by a variable brake of **conservatism in the patrons** of invention, whether enterprisers or consumers. The more restricted, select, specialized and intelligent is this patron class, the less will be its conservatism and the faster progress.

30. The popular notions of great inventors are essentially mythology; and that man is preferred as the **titular inventor** who belongs to one's own or a related country, and who first achieved commercial success with the device.

31. The craft of invention reveals the following **trends**: (a) from the empiric toward the theoretical; (b) from the unconscious inventor thru the amateur to the professional one (c) from the evolutionary toward the discrete or broken, epochal; but (d) from the accidental toward the deliberate and sure; and (e) from the individual source to the organized inventing group. (f) A consequence of these trends (and others) is a prodigious increase in the efficiency and rate of invention.

32. The inventions which revolutionize a device or industry are commonly made by men **outsiders** to it yet informed regarding it; the far greater and more valuable mass of perfecting inventions are made by **insiders.**

33. The inventors are in **partnership** usually **with enterprisers,** whose courage, intelligence, business sense and wealth are commonly of importance comparable with that of the inventors themselves.

G. Effects of Invention

34. **Equivalent invention:** perceived needs are met by various **unlike,** as well as duplicate solutions, so that any great invention is simultaneously paralleled by other, often utterly dissimilar means for reaching the same end at the same time, e.g., reaching California by clipper, steamer, pony express, railroad and telegraph. Inventions may be seen as arriving in functional groups.

35. Hence **no** single **invention** ever **revolutionizes civilization,** nor brings, simply thru having been invented, any important changes in the life of the mass of men.

36. An invention coming **before its time** remains undeveloped and practically useless.

37. The **desirability** and acceptance of inventions are matters of their meeting the felt needs of certain individuals, and not of society; so an invention may flourish and be practically imposed on all, thru competition, tho it only decrease the general wealth or welfare.

38. (a) Inventions may save labor, land, capital, lives, or may not save anything, but add science to destruction, or bring a private profit, or a new enjoyment or convenience for the public. (b) Between **labor—, land—and capital-saving** inventions the discrimination of effects is complicated, but as a rule the factor chiefly economized loses in relative share of the distribution of the national income, while usually all ultimately gain in their absolute shares.

Two of these Principles, the 13th and 30th, we have extensively discust in *Inventing the Ship,* respectively in connexion with the beauty of the sailship, and the Fulton and Ericsson myths, so we shall consider them but little in this volume.

CHAPTER II

EXAMINING THE NATURE OF INVENTION

Continuity of Development (Prin. 1)

Of the Principle (1a) of continuous evolution by petty accretion we have incessantly presented instances in our companion volume,[6] but we should not leave it without several other viewings, so vital is it to all understanding of invention, and yet so commonly disbelieved.

The Biologic Analogy and Factor (Prin. 1d)

We have said that the evolution of the ship much resembles biologic processes. Many others have said as much, e.g., the nautical authority H. H. Brindley says:[7] "The resemblance between the problems of nautical research and of zoölogy is a close one. Change of function by an organ and useless vestiges are everyday trade of the comparative anatomist, and the history of sailing craft also furnishes plenty of them. The mind of man working collectively thru the ages has brought about the same effects on his inanimate products as nature has evolved in the development of species. Ships are a parallel with the pottery of primitive man and the latest types of express locomotive; both exhibit, *pace* the engineer, 'useless vestiges'."

Is the biologic analogy anything more than an analogy? Yes. The ship or any invention *is* a biologic organ, in the same sense that a bird's nest is, if not a snail-shell. Each is something made by a living creature, in order to meet the needs of his life, and it has grown up, evolved, along with the fleshy parts of him, by a long process of mutual adaptation. The young birds have become so delicate that they cannot live without a nest; the nest has become so well built that it strongly promotes the survival of its bird, and has become almost as much a part of him as his bill and feathers. As the apes use sticks and stones, so primeval man used chipt stones, and later evolving man has used ever more of tools, ships and all inventions. These have so intensified and multiplied his life that he lives grandly and long, and has spread over almost every land and into the very waters and air, and has so swarmed in numbers that his total of body weight exceeds that of any other species, plant or animal. Always his tools, his inventions, are essential parts of his biologic life. And he has gained them thru evolution as truly biologic as that which gave him his manly beard, his unparalleled capacity for mating activities, his large, vaulted cranium. For he has gained those inventions thru modifications within that cranium—a growth of reasoning power, a decline, probably, in definite prescriptiveness of instinct, and a growth of those qualities like will and attention—all of which both enable and drive modern man to *invent* and make endless clever efforts to better his ways of living, following psychological patterns.

We cannot, indeed, be sure that this evolution in native inventiveness has continued, parallel with invention, during historic times; or even since the Crô-Magnon race. Most writers assert it has stopt; but all the *evidence* inclines us to suppose it has continued until the last century, or until Monasticism or the Reformation, when the abler classes of men ceased to have larger and often plural families, and our race divorced Eugenics for the real adulteress, Dysgenics.

Insofar as evolutionary selection takes place not within but between nations, there has no doubt been a biologic growth of invention until almost or quite today. Ellsworth Huntington points out[8] that almost every important invention has increased the population of the tribe or strain able to contrive or early to appropriate that invention, especially when it was a means, such as fire and warm clothing, for taking and populating desert or little peopled land. So the northern border of man's originally subtropical habitat must have been successively peopled, one colder zone after another, by the most inventive or adaptive people of a stock previously selected by the same principle—*crême de la crême.* And so we have the particularly inventive New Zealanders, Australians and Yankees, latest offshoots of the great northern British race, filling new continents with their millions, largely because Britons were leaders in the invention and confident use of *ships.* (Of course we do not wish to deny the many cultural, geographic and historical factors involved in this question, or any other; but we are contemplating here the biologic only. The modern Britons possess inventive ability, and those other factors make it express itself a-voyaging.)[9]

However it be about modern evolution, it is certain that invention and all its products rest upon man's brain as closely as do his scalp or *pia mater;* that every turn of invention must be determined (conditioned) by man's biologic inheritance (as well as going in turn to modify that thru conditioning racial evolution). Now hence come those striking similarities between strictly biologic and technic evolution. If there be Martians, creatures of wholly different ancestry from ourselves, perhaps they

[9] Not only has invention been pushing northward the frontier of human habitation, expanding the domain of the most advanced races, and helping the more inventive nations and (until recently) families to prevail in competition over the less inventive, but invention has likewise been shifting northward and toward various desiderata the scene in which civilization will *most* flourish, as I have shown in the theory of coldward progress, N 291. The "ridge isotherm of civilization" is today that of 50°F, almost exactly the mean annual temperature of N. Y., London, Paris and Vienna. Each shift of civilization poses new problems of adaptation. And it is closely along that isotherm that almost all the world's inventions are made today. Cf. ftN *43,* N 194 and Prins. 16 and 6.

think unfamiliar thoughts as readily as familiar, contrive great inventions overnight, and all adopt at once every improvement. But Terrestrial inventing must be done by the animal *homo sapiens* (fond misnomer), a poor creature whose brain, and therefore mental products, simply do not work that way, any more than those of his cousins the spiders and beavers. As we are and think, so do we build and devise. This is the source of the whole biologic characteristic of inventions.

We may now answer the question of whether it be legitimate to speak of the evolution of the ship at all. MacIver says [10] that if evolution be absent from a series of changes, the matter is for the historian, not the sociologist; and that evolution "is an order of change which unfolds the variety of aspects belonging to the nature of the changing object, in which potentialities lying within it are made actual." The forces changing it must not come all from without, and the change must permeate the whole character of the object. . . . It would accordingly seem legitimate to speak of the evolution of the ship, if we reflect that that invention, the ship, is not just a lifeless thing of wood and metal, but a vast complex (Prin. 2a) of which most elements are social thoughts of man, a set of social institutions, customs and ideas about building and using ships, the whole associated together and named by custom.[11]

Stagnation and Progress

Let us now review our nautical history, asking first in what epochs chiefly the ship got invented, and later where, how, and by whom and why, in order to reach what inductive laws we may of invention. We shall be struck first of all with the fact that inventing seems to have been done with very different frequencies, rare and short periods of progress alternating with ages of near stagnation

[11] Cf. our later definition of a steam engine, by a sociologist (p. 26).

in invention. Then we shall note that when invention flagged, *improvement* did not, unless the general level of civilization fell; the ship became better, even much better *without* any noteworthy inventions.[12]

At first there was, apparently, very slow development, with what inventions were made falling in the ages and countries of advancing civilization. The progressive ages, for the merchant ship, were first during the rise of Egyptian culture to the Pyramid Age, and then the 1st century of the Roman Empire, which greeted the topsail, *mendicium,* mizzen and paddle-wheel log. After that there was stagnation or retrogression until the finding of the lateen. Much later the Mediterranean galley was almost changeless for five centuries, until its extinction ca. 1800. The Viking ship seems to have been changeless for several centuries, then was profoundly modified in the 1000s, in form and by decks, but not in mechanism, thru Mediterranean influence. The resultant "cog" was rather stable again until the 1400s, when another period of instability and Mediterranean influence thoroly modified her appearance, in the directions of height, intricacy and better sailing qualities, but still without any notable inventions. A century and a half of slower progress followed; then from 1650 to the present day has again been a period of notable stability for the sailship, especially in contrast to the tremendous inventions in other branches of civilization. In 3¾ centuries there is not an invention adopted in the full-rigged ship, more important than stuns'ls, staysails, the spanker, steering-wheel and auxiliary engines. Jibs, iron and steel were borrowed from other vessels. From 1720 to 1870 there was a particularly uninventive stretch, yet marked by great progress in size, speed, general efficiency, science and beauty. Only today, faced by the imminent extinction of all wind-blown vessels, do we see signs of an impending revolutionary period—such signs as the auxiliary engine sailer, the sailing barge, and especially since the War, the metal sail with two stream-lined surfaces, and the rotorship. We are again reminded of our biologic parallel, wherein evolution proceeds by far the most rapidly in those rare periods of climatic stress, which massacre old types and seem to breed new ones which can meet the new conditions, like the glacial age that produced man. Mutation or extinction—those are grim Nature's alternatives when conditions change and strong rivals appear. It needs no wit to see, that the winged ships today are facing a similar alternative. It is sad that such beauty must be exterminated, or degraded to the unlicked ugliness of novel forms. But technologic evolution is as inexorable and ruthless as the biologic, because it is an outgrowth from it—men compete with men today not by teeth but by tools, not by thews but by thots.[13]

In the history of the steamer the rate of invention has been likewise variable. At first we had slow progress and few boats built, because those were all failures. Then experienced engineers took hold—Symington, the Stevenses and Fulton—whereupon progress became faster, and far faster after large commercial success. Then about 1880 or '90, after the introduction of triple expansion, progress seemed to stop, at least so far as concerns the large features of the freighters which are the chief ancestors of the rotorship. But in other lines, as in the express steamer, warships and motorship, and freak freighter hulls (the whaleback, turret ship, blister and Maier forms) progress continued unchecked, likewise in minor details.

[13] Chapin points out that there is a long-run natural selection between inventions of economic or military significance, in the rise or decline of the peoples adopting or rejecting them. N 152. Cf. also N 7.

Why the Sporadic Distribution?

If now we seek the cause of the variations in frequency of invention, our first procedure should be to seek a correlation, mental if not mathematical, between changes in the rate of invention and in any other variable that we suspect is related. Suppose we take height of general civilization, to compare with the frequency of invention. The peaks of marine inv. activity were say in the years 1, 1050, 1450, 1850 and 1925. There results no correlation with the progress of civilization, except the general one, by which every kind of progress shows a speeding up in the course of the ages. There is no more agreement between the particular dates of inventive activity and cultural progress than mere chance would dictate. Even less correlation appears between these inventions and any other single factor that the writer can think of.

Suppose now we contemplate not the ship's noteworthy inventions, but her general progress, signified by greater speed, seaworthiness, size, fighting power, durability and any and every kind of efficiency. That is to say, suppose we consider not only the major inventions recounted in our published history, but also the thousands and millions of minor inventions, little improvements, modifications perhaps so slight as to be unconscious, in dimensions, proportions, form, substance or use, and small wise thoughts such as adding another sail like the previous ones. We now find our curve of progress straightening out into a smooth, constant rise since the Dark Ages, whose slope of ascent is parallel to the slope or rate of general progress of civilization. And the line projected backward from the Dark Ages shows the same close correspondence to the ups and downs of civilization, so far as the history of the ship is known. Similarly in Graph 3, the course of all German shipbuilding patents together is much smoother and better correlated with civilization than the graphs of particular marine inventions. This exemplifies the familiar statistical principle that increase in numbers of items brings stabilization of rate.[14] Still smoother are our curves elsewhere [15] of the rise of steam pressures and decrease of coal requirement, which reflect many inventions, plus minor betterments, plus the absence of adjacent years.

Revealing "Invention" as a Matter of Language (Prin. 1b,c)

So general efficiency correlates with civilization far better than do major inventions. Is not this the explanation?: The great truth, the essential process, of invention and technic progress, is reflected under the head of general efficiency, not of major inventions. The multitudinous little turnings-up not only overbalance the big in importance, but include the big and compose them. For what we call a major invention, like an efficient fore-and-aft rig (9 centuries in the making), rudder, steering wheel, steamship, or a propeller or oil engine worth using, is composed of at least a dozen minor inventions and usually of thousands, all of which are bound together by some physical principle that they all involve, and by a thread of *language*. Our English language provides a single word "steamship;" hence we must count the beginning of the steamship as one great invention. Our language does not offer a single word for a "*ship whose hull is designed with theoretic science for minimal resistance to forward motion compatible with easy construc-*

14 The dashed curved lines near the bottom of the chart signify that there the patents fell to zero, which cannot be plotted on a semi-logarithmic or "ratio chart" like this one. This form of ruling, logarithmic in the vertical (patents) scale, was adopted because it shows equal proportional increases by equal rises. Thus a straight upward sloping line would indicate an increase at a constant rate. The data were collected for me by Dr Barkev S. Sanders, from Patent Office records . . . Cf. Kroeber, ftN *69*. Historical tonnage stat. are in Kuznets, N 166, pp. 131. 426-8.

tion—dry-docking, etc.'' Hence the origin of the latter ship is not to be listed among inventions, like the steamship. Yet by every consideration save language the two new ships are near a parity. If we test by Physics, they have equally a unifying principle—use of steam power in one case, and of hydrodynamics in the other. If the test be Economic, each had its chief development within the same limited period of years. If our names for things were but different, as they may be in Arabic, we might never have heard of ''steamships'' being invented, but only of steam engines, and ships using them, as some ships use derricks, some elevators, some electrical equipment. On the other hand, we might be speaking of the invention of the steelship, the wave-straddling-ship, the powership, the weldedship, the manygreatsaloons, the intraspecializedcrew. If we do not find such words in any of the languages with which we are familiar, it is because the constant mutual influence between the Western European tongues has made them so much alike that their vocabularies often seem but dialects of an international language (a fact which Esperanto turns to good account to form an easily understood, neutral tongue). In marine matters English has been especially influential. Thus the world no longer speaks of a ''boat made to go by fire,'' nor of its old Greco-French equivalent *pyroscaphe* (fire-boat—Ital., Sp. piroscafo), but today of steamboat, bateau à vapeur, battello a vapore, barco de vapor, Dampfboot, parovoye sudno, all of which are simply ''steam boat'' translated literally, and are quite inappropriate, since the vessel is no longer a boat but a ship.

Defining Inventions—the Difficulty and Statistical Need

Well here we are in the midst of an attempt to define invention, when we have intended to write our book without a single complete definition in it, but only partial definitions to avert some particular ambiguity. Definitions are hard things to explain and impress upon the reader, or even the writer, and hardest of all to impress upon the language, changing the practice of other writers and millions of speakers. And of all words in the language perhaps the hardest to define is "invention." Because invention is the only thing that is never twice alike —by hypothesis each invention is different from each other, and was novel and unforeseen. A treatise on patents[16] gives 45 pages to explaining the word "invented." But I prefer the dictum of the Supreme Court: "The truth is, the word cannot be defined in such manner as to afford any substantial aid in determining whether a particular device involves an exercise of the inventive faculty or not."[17]

Yet we are not willing to dismiss invention as quite impossible to define or measure; for that would be to discredit all statistics of invention. We have already been using some informal counts, and should like to use more statistics, as have other students of the subject. Statistics begin with counting, and counting begins with defining the unit to be counted. But if that cannot be defined—? A patent, now, is another matter. Patents are granted in great numbers and their statistics compiled, by patent officials, in accord with elaborate laws, rules and customs, in return for fixed fees, all of which make for sufficient uniformity for somewhat shaky statistical purposes, especially in a single patent office in the same field and age. (When we compare the grants of different countries andor periods, statistical precautions are called for that are not always used.)[194] Patent statistics have many uses, but cannot be extended far into the past, and being very voluminous, organized only by fields, and their items mostly rubbish, they cannot well be used for statistics of great inventions, or for counting and comparing those big aggregations, each of thousands of detail inventions, which always constitute the great and famous "inventions" in which people are commonly interested, and can discuss with shared acquaintance. Something, to be sure, may be learned from statistics of patents in particular fields, like propellers, sails and rudders, as in Graph 3.

But when we leave the apparently neat statistics of patents, for what reason forebodes as the vasty, undivided continuum of inventional reality, presenting no countable divisions save those imposed by language, are we not in the position of one who would count waterfalls? A waterfall seems a definite thing, as does some invention like the steam engine. But under the light of reason all totals of "waterfalls" would dissolve like their own spray; because there is no possible way to draw dividing lines between waterfalls vs. cataracts, or vs. trickles of water, unless by arbitrary measurement limits, such as are not possible with invention, or by arbitrary language usage.

And yet one might say "*some* things are *certainly* waterfalls, and we may consider these alone. Similarly, are not some things certainly to be clast as 'important inventions,' and could we not base some rough statistics on such only?" Very well, let us take what is probably the greatest invention on the ship, and one with which we are all perfectly familiar, the steam engine; and let us see how this invention might be defined, delimited and seen by half a dozen competent people. In our little symposium speaks first the

PHYSICIST: A steam engine is any device which uses the expansion andor condensation of steam to get from random to organized molecular motion. Its most important form today is the turbine; but an internal combustion engine is a steam engine in part, because it depends partly on the expansion of water vapor, as well as using a mechanism essentially the same as that of a reciprocating steam engine.

Mechanical Engineer: A steam engine need have no steam at all. It is obviously the same invention in those forms where mercury, air or some other fluid is substituted for water. Moreover a liquefying pump is the same thing, with its working reversed; and a gas engine is the same mechanism, only with the combustion internal. But the steam engine does not include the steam turbine—that is a basicly different device, depending on the impact of a swift jet in an open chamber, whereas steam engines press out a slowly sliding wall of a closed chamber.

Metallurgist: It is obvious that a steam engine must have steam, and that a turbine is as much a steam engine as any other. The steam engine was never invented—it is constantly being invented today. And this invention of the steam engine is and has been mostly a matter of metallurgy. The early steamboat builders used almost zero pressure, simply because they knew their coarsely built wrought iron boilers and other parts wouldn't stand higher pressure. Today we have high pressures, turbines, and tomorrow perhaps internal combustion steam turbines, chiefly because metallurgists provide suitable alloys and shape them with hair-breadth accuracy, to meet the fearful punishments from heat, strain and vibration which the new engines impose. When metal parts equal to such demands have been created, for diesels and many other uses beside steam engines, then the mechanical engineers can easily devize new engines to utilize the new powers. *Their* work is only secondary; the heart, the chief evolution control of the steam engine is its metals and their shaping.[18]

[18] Dr. Wm. Kerr, M. I. Mech. Eng., says of the gas turbine: "It is well to realize, however, that its successful development depends less upon fortuitous invention than upon the provision of all the resources of systematic investigation and the progress of metallurgy. The same could, perhaps, be said of all the probable lines of development with which the engineer is confronted." *Discovery*, 10: 407, 1929.

LUBRICATION ENGINEER: The steam engine is not an invention—it is at least six inventions, to wit, the boiler (fire-tube or water-tube), the reciprocating or turbine mechanism, the condenser when used, and the internal lubricant. To these may be added an indefinite number of further inventions, as the safety-valve, injector, superheater, stoker, pulverized and oil fuel, etc. People commonly overlook the fact that progress in all these inventions has been largely due to finding better lubricants, which enable higher temperature and pressure and faster working. The engine of the future, the internal combustion turbine, is largely held up on this account. Low temperature ones exist today; give us but a fit lubricant and your engine will come automatically, along with much other useful machinery.

STEAM ENGINEER: The steam engine, the power that made modern civilization, includes the boiler but not the stoker nor fuel, nor water-softening nor purifying nor distilling devices, beyond the first 3″ of the feed pipe; but the injector is included, and the safety-valve but not the whistle. Turbines are included, but not internal combustion engines. The steam engine is a branch of Machinery, in Engineering; fortunately Sociology has nothing to do with it.

SOCIOLOGIST: "The steam engine" is not an engineering but a sociological phenomenon, a complex of folkways, and of thoughts which are socially inspired at every turn. If every steam engine in the world were destroyed, "the steam engine" would survive, and at once be built again, because the essential ideas, of how to make it and what for, would persist in people's minds and on records. Engineering explains the physical principles, but only Sociology and Psychology can explain the invention, i.e. how the thing comes to be invented. Also, an invention has probably no existence, i.e. definition, at all, except what is imposed upon the engineering objects by the standardizing tendencies of social man.

Now gentle reader, if you still feel inclined, you may add your own definition of the steam engine. But be sure you're right. We are not concocting definitions of "an invention" for the fun of it; we are seeking to get a sounder idea of what an invention really is. Suppose we try to define that great invention in the ship, the propeller.* Its most essential element is blades swept thru a fluid at a slight oblique angle, in order to obtain a thrust forward, i.e. at right angles to the direction of motion, in maximum proportion to the thrust in line of motion. This is identical with the blade of a sculling oar, some rotary fans and pumps, a windmill* (reversing the action does not change the principle), an airplane's or autogyro's wing* as well as propeller,* a kite, paravane, most steam turbines, some water-wheels*,[19] a contrapropeller and other deflecting blades, a surface or a ladder hydroplane, skipping stone, boomerang, windvane,* rudder,* stabilizer,* feathering of an arrow, deadwood, and a sailship* close-hauled. It is the same as a fish's tail too, and related to the inclined plane, wedge and screw for solids. The various inventions named first have each their varying elements in conjunction with the oblique blade principle, making them novel combinations; but yet when Flettner substituted Magnus' rotor for the blade, he was allowed a single patent[20] with 55 claims, covering by name all the inventions starred above, and probably others in principle, under American law which allows only one invention to a patent. In primitive times when some of the

[19] Non-centrifugal ones having the axis parallel to the stream-flow.

[20] On an "arrangement for exchanging energy between a current and a body therein." U. S. reissue pat. No. 18,122, issued July 7, 1931.

simplest uses of the oblique blade were invented, they were triumphs of intellect. A century ago the harder ones were still inventions. Today a new use for the oblique blade would not be even an invention, but only the application of a well known principle of engineering. If we could concoct a definition of invention, setting a lower limit for inventiveness, it would have to change to suit each different age of industry, demanding more from the generations that have more. This illustrates our PRINCIPLE 5, as would also a glance at the simplicity of the earlier inventions[21] in the history of the ship.

If we attempt as above to define an invention as the utilization of a single physical or chemical principle, which might have some inherent individuality, we are commonly led into such difficulties as with the oblique blade, where the definition would oblige us to lump as one invention what custom stoutly declares to be different. There would also be question of which of the various principles involved we should take as our criterion of the invention. One example we have seen in the steam engine. For another, the close-hauled sailing ship might have been taken not as an example of the oblique blade, but as the same invention as the same ship running before the wind, which is the same aërodynamicly as the parachute and sea-anchor, which are the aërodynamic opposite of the close-hauled ship. An invention must be defined as a combination of several ideas; and which ones are involved, and which are more important, is always a question. For instance, one of the best writers in our field, defines the automobile as the six-element combination patented by that sharper in ambush, Selden.[22] Yet the courts finally decided against him, and one could build

[21] "A century ago many of these modern mechanisms would have been hailed as an epoch-making invention, comparable with the greatest ever conceived by the human intellect. Today we accept them as a matter of course. Invention has become a profession, a branch of engineering." W. Kaempffert, N 199, p. 36.

what could be called an automobile without using any one of those six elements, save running gear.

This PRINCIPLE 3a, that invention is a new combination of old ideas, will need little proof, nor will its further claim (b), that the number of elements may be indefinitely great, and not just two as Tarde claimed, unless in some psychological sense that need not concern us in Sociology. Chapin says[23] that invention is most favored when all the elements enter a zone of potential integration, where all are scientificly understood. Those we are most apt to think about are those that but recently advance. (cf. also Prin. 7) The inventor of the ladder hydroplane took cheap steel for granted; yet that element has existed only since the 60s. This craft is held back especially by the danger from floating driftwood in harbors, a problem susceptible of either social or mechanical control.

FINDING FAIRLY SOLID, SOCIAL BASES FOR DEFINITION

Shall we conclude, then, that "an invention" is only a quite arbitrarily and variably delimited aspect of an endless *n*-dimensional continuum of technic constructions and knowledge which is the *real* situation, and that the only proper name for an invention is Bottom's Dream, because in very sooth it hath no bottom?[24]

No. For in this dissolving invention into a mist of language we have at last got to something solid. For language usage is an objective phenomenon, and a widely social and therefore somewhat stable phenomenon. That is to say, common words do not change their meanings suddenly, since they represent the choices of so many mutually inter-influencing people. Furthermore, it is familiar to statisticians that at any given time the nature of almost any social phenomenon, when measured according to most criteria, is found to have a frequency distribution following more or less the bell-shaped, normal-frequency curve, or at any rate unimodal, i.e. having a central high point about which the cases are largely grouped. One would therefore expect, for instance, that if one took all the definitions that English-speaking people would give of "propeller," and a picture of a modern steamship in longitudinal section, and if one drew a line along the tail shaft for the distance included in each definition, the lines would largely coïncide, and be most numerous in the region of the propeller proper, notwithstanding that some would extend the definition aft to include the contrapropeller and the fairing of the tailshaft into the rudder post, and some would extend it forward to include more or less of the shaft, gearing and engine, remembering how greatly the engine had to be modified to fit it to screw propulsion. Despite these extreme definitions, the clustering of voices around the screw proper would suffice to give a fairly uniform definition of this invention, and a distinctive center of significance, objectively and sharply differentiating the propeller from all other inventions that would be represented by positions along the tail shaft. Likewise for its traits that might be represented by points in other lines, such as inclusion of propeller-building methods, and of hydro- and aëro-dynamics, it seems likely that the same method would yield a definite identification, a single mode. But for other questions, such as whether "the propeller" can be an air screw, and whether "the steam engine" includes the boiler and a turbine, our method yields a bimodal result, having no meaning except that of a majority vote among a group selected to answer, a sort of definition little acceptable to a scientist. But by and large it seems that common language usage does furnish passably satisfactory definitions for the greater inventions, sufficiently uniform, consistent with each other, and objective to the user, to justify the social scientist in talking about inventions, and even using a little simple statistics on these, provided little accuracy be claimed for them. We can say to some extent "Est modus in rebus, sunt certi denique fines."

Language is assisted to this uniformity by the social unification in the habits of thought which are behind language. Everyone thinks of a propeller as entirely different from a rudder, notwithstanding the identity of their vane principle, and the fact that either can after a fashion replace the other. Likewise a ship's and an airplane's propeller are commonly thought of as the same, while a windmill is in a different world of thought, altho a philosophical physicist might hold that an air propeller, whether in a blower or an aircraft, is identical with a windmill, as their best shapes also are.

And finally language is assisted to uniformity by our industrial habits. Constructive man groups and standardizes his products, instead of commonly building everything possible. We do have some steam engines acting without steam, or without fire, but they are very few. We can make usually good distinctions between a motorboat and an airplane, because until recently there have been no intermediate types built or imagined, altho we are beginning to get some, in the hydroairplane, the ladder hydroplane, and a recent type which rises entirely clear of the water, yet cannot leave the cushion of air next the water. Hydroplane or airplane? But anyway it is scarce, and many of the possible links between inventions are entirely unbuilt; so our common definitions can persist.

We shall continue therefore to talk about and even count inventions, altho it seems doubtful that they have any existence at all, i.e. any boundaries, aside from the social conventions of language and thot, the customs of industry, and the fact that often the intermediate forms, which would make one invention merge into another, have not yet been invented, or if thot of are never or seldom built. We shall try to educe more of truth by juggling conventional errors, hoping that as the reader turns the pages of our book he will grow in understanding of the subject, and make better guesses as to wise public policy, even if we cannot yet, or perhaps ever, hope for a positive, accurate, quantitative grip upon any part of the subject . . . . Our defense is first of all *tu quoque,* which is a poor argument, but a first class silencer. Who in Sociology does any better? We talk about leadership, Episcopalianism, criminality, the Industrial Revolution, when no one ever has defined those words, or set any definite limits to the phenomena they include and exclude, unless it be a legalistic definition, like formal membership in a church or a prison, which has little real meaning, corresponding to sociologic comprehension of the degrees of fact and causal relations. Our excuses are the English language, and the concepts provided by the folkways of modern thot. A hodge-podge from undirected evolutions, our speech is as full of misconceptions and vagueness as of words. Yet we go on using it—have to—it's the only means of communication we've got. We must ask like Shakespear's prologist,[25]

> "may we cram
> Within this wooden O the very casques
> That did affright the air at Agincourt? . . . .
> Where—Oh for pity!—we shall much disgrace
> With four or five most vile and ragged foils,
> Right ill-disposed in brawl ridiculous,
> The name of Agincourt. Yet sit and see,
> Minding true things by what their mockeries be."

Do Inventions Follow a Life Cycle?

That important inventions, however arbitrarily defined, do have some inherent existence, is suggested by the considerable number of Principles, three dozen or more, which we and other writers have found to hold true of inventions, with rather few exceptions. A Principle which seems especially appropriate to take up here, because if established its mathematical nature might give some further indication of the reality of "an invention," is our Principle 12, that inventions are rarely rejuvenated. Tarde said that the use of an invention shows "a slow advance in the beginning, followed by rapid and uniformly accelerated progress, followed again by progress that continues to slacken until it finally stops."[26] A fourth stage of actual decline would be often added. Kuznets [166] suggests a similar cycle for inventions, and Chapin adds that the pattern is first an attempt to use the old and a slow growth of the new; then rapid adoption of new features; then last a slowing up, and an integration of the new into a rather fixed system.

These rules seem to hold true very well; but let us see how much they mean. First we may dismiss the claim for a slow start, as probably a statistical illusion, due to graphing the progress of the invention on an arithmetic scale. When nine of Kuznets' counts by year-groups of perfecting patents on certain inventions are graphed on semi-logarithmic, ratio-chart paper, which is more valid for present purposes, since it shows equal proportional changes by equal heights, the slow start disappears in every case. (But a slow start is probably a real tendency—Principle 8—if one go back to the pre-commercial period.)

Suppose we ask a man how the stock market will go tomorrow, and he answers that it will continue rising, unless it has reached the phase of stability, or perhaps the period of decline. The man knows nothing about it, and has told us nothing. His prediction is meaningless, so long as it allows for all conceivable events. The omission

---

[26] Continuing, "These then are the three ages of those real social beings which I call inventions or discoveries. None of them is exempt from this experience." *Laws of Imitation*, N 176, p. 127.

of the phase of stability, however short, is not possible, according to an axiom of social science that social changes never come about with absolute suddenness, but could be graphed in curves without angles. To have a real law on inventions we must predict something *less* than the totality of possibilities, e. g. by placing time limits on the duration of those stages, as no one has nor shall we attempt, or by propounding a rule that the cycle is never repeated. "What, never? . . . . . 'Well, hardly ever'." That is our Principle 12, that inventions are rarely revived, the increasing use of an invention tends to halt only once, or in still other words, its historical graph rarely shows two peaks, nor a plateau followed by a renewed rise, but it might show a plateau followed by a decline, or continuing into the future.

To test the law we have glanced thru our history of the ship, and found only eight cases of possible exception to the rule, viz. the ram, reef-points and bonnets, a second and a fourth mast, topsail, leg-o'mutton sail, iron ship, and balancing the rudder. In half of these cases, however, the revived invention is essentially rather different from the old one, as steel clinker building differs from wooden, modern semi-balanced rudders from the paddle quarter rudder, and the ship of iron puddled by machinery, from hand puddling. Since the number of inventions included in our mental survey was probably a couple of hundred or so, the observed frequency of recrudescence after a decline seems smaller than would be expected from chance in the absence of a tendency. Cursory consideration of other than marine inventions indicates the same rule. While we often hear of some old invention being revived, one almost always finds that it is not a clear-cut case. Either the invention was hardly used at one epoch or the other, or the new form is quite different from the old. The reason is evident—that if an invention really fits into a civilization, so that it is freely used, it is not apt to fit into a widely different civilization. Think of reviving any of the disused ancient inventions and you will at once appreciate this point.

We may even, from our historical evidence, hazard our PRINCIPLE 9, that the *rate of increase* in the use of an invention probably tends to decline continually from the start of its practical success, and in time will become negative, i. e. a growing rate of decrease. The various consequences of this thesis may be illustrated by the four curves of Graph 1, which assume an invented device of considerable durability, and which lay no claim to mathematical correctness. Starting with curve No. 1, a constant decline in proportion of new construction to that extant, we see by (2) that the cumulative absolute adoptions increase toward a limit, while (3) the net absolute increase first climbs rapidly, then falls below zero from the removals of wornout devices, and last becomes insignificant. Curve 4 shows how the number of devices in use rises to a later peak and then approaches zero.

Statistics on the utilization of 82 inventions begun or flowering in the last generation, and having great social significance, gathered by the writer for Professor Ogburn,[27] give no clear evidence on changes in adoption, because the period covered was only about 8 years. All but 15% were advancing in utilization or construction, and the median of all rates was an 11% growth yearly. For 30 inventions whose measure of direct utilization was printed in the report the geometric mean was +19.2% yearly, if it be permissible to state an average in inventions, while the 11 cases where the count was of the production of a durable capital good, had a geometric mean a little lower, +17.6% yearly.

The high frequency with which our not inevitable mathematical principles hold true of invention (Prins. 12 & 9, of only one peak of use, and continual decline of rate of increase of adoption) do suggest that there is some intrinsic reality and unity to the concept of invention. But still this characteristic of consistent reality may have been merely imposed upon the subject by man's linguistic and other standardizing tendencies.

Inventions that Cease Evolving

The rise of an invention may be measured by the statistics of its utilization, such as we were considering; or by observing the number of accessory, detailed inventions which are added unto it. The two measurements might be expected to follow along with each other, as Tisell [28] and our Principle 15 claim. But the data which he and Kuznets [166] present, and our own in Graph 3, show very numerous discrepancies, as when the still-flourishing acetylene light [29] was forsaken by inventors to follow a new and more attractive lead. And Kuznets shows repeated cases of inventions, such as the plow and electric motors, meters and cables, whose perfected patents declined rapidly while the use of the invention was even increased. The graphs of accessory patents reveal second peaks and apparently every possible shape. But to be sure, they refer to shorter periods than we have been considering. In the history of the ship we recall a number of cases of a long inventional standstill, without going back to uninventive centuries when stability was to be expected. Some cases ended in the disappearance of the invention, or seem so ending—the square sail and full-rigged ship, steeple engine, and walking-beam engine with Stevens' valve gear, a peculiarly unchanging invention. In more cases the standstill was followed by further evolution—the quarter, ordinary wooden and plate rudders, the rib, muzzle-loading cannon and mount (part of the merchantman, and fairly stable), Admiralty anchor, and the Mississippi packet-boat. In a few cases the stasis continues with the future uncertain—for clinker building, the ordinary and feathering paddle wheels, propeller, triple expansion marine engine and the stockless anchor.

In fine, Principle 15, that the perfecting inventions made on a device tend to vary with its absolute extent of production, while it seems inevitable from logic and is upheld in general by observation, has many exceptions in practice, due to the operation of the very numerous forces controlling inventions. A striking type of these abnormalties is a period of stasis in design, ending sometimes in the death of the invention, but more often in renewed evolution, or the stasis not ending as yet.

Why do Inventions Seem to Decay?

Kuznets says, "The general impression, however, is that in manufacturing industries in which the so-called industrial revolution occurred long ago, the rate of industrial progress has diminished with the passage of time. The industrial arts in their technical aspects are gradually approaching stable forms."[30] This is for industries with an unlimited supply of raw materials, and is not claimed to be proved, but further investigation to be called for. Particularly the early step of introducing mechanical power has an importance, he thinks, which is never equalled during later progress.

Is all this true and does it apply to the ship and to inventions in general? Certainly we have not found any tendency for marine progress to decelerate, but the reverse. The introduction of mechanical power took place in prehistoric ages, first water power in descending rivers, and then the sail around 6000 B. C. Perhaps indeed no later invention has equaled these in importance to shipping. The later introductions of power, for mechanical propulsion, steering, weighing anchor, and general hoisting and pumping, came almost automaticly when they were called for by the ship's growing faster than her crew, except for the steamboat they made no particular stir, and have continued, it seems to the writer, to evolve at a good rate, as has ship propulsion. For the ship, then, no evidence for the principle holding. But does it hold in general?

Julius Wolf has propounded four "Laws of Retardation of Progress," including "Every technical improvement, by lowering costs and by perfecting the utilization of raw materials and of power, bars the way to further progress. There is less left to improve, and this narrowing of possibilities results in a slackening and complete cessation of technical development in a number of fields."[31] Kuznets and Chapin support this, but it seems obscure and needing restatement as our Principle 10, which see. Inventions lessen their own importance and thereby check their further perfecting only when they bring an actual reduction in the amount of labor and other productive forces employed in their field. Whether this reduction will happen or not depends on the balance of opposing pulls of considerations relating to the final good produced and the supply of raw material. These are *flexibility* in ultimate demand and natural supply, tending to permit expansion of the use of the invention; and *proportionate share* of the invention in the total productive process, the greater the share, the greater being the reduction in total costs.

For example, in the navigation of the Mississippi, the addition of steam power to the previously used poor winds and one-way stream power, brought such a vast improvement in the total process of living along the Ohio (whence goods had to be shipt out or in thru New Orleans, life in the West being largely a struggle for transport), that population grew fast and likely the proportion employed on the river and in shipbuilding increased, despite the fact that one steamboat would replace 100 keel-boats. There was no limit of natural supplies, and the demand for transportation was highly flexible. The packetboat became a great and famous part of midwestern life. Various inventions improved them, although a period of marked stability ensued after 1840. Somewhat later, when men learned to replace the fancy packet by a mere tug pushing cheap barges, the Mississippi steamboat dwindled, passed from men's thots and became notably unprogressive until the modern (waterways) era, notwithstanding that as much freight continues to be carried on the Mississippi as ever. The barge and new inventions cut the transportation cost to be the lowest on earth, around a mill per ton-mile, but even if they lowered it to zero, they would still not increase the volume of river traffic greatly, because riverwise freight costs are today but a small fraction of the total cost of making and marketing goods in the Mississippi Valley. The earlier steamboat inventions built up, and the later ones tore down the river transport business, just from the considerations of changing flexibility of demand for transportation, and the differing share of river freight charges in the total cost of production.

Similarly the old importance of physicians and bedside medicine has been cut down by the invention of the telephone, automobile and more hospitals and city living, because these double or treble the efficiency of the doctors, in a profession with somewhat inflexible demand. Its demand would be more flexible if healing had good substitutes. Commercial aviation is a business with good substitutes, so any economizing invention that makes a *large* improvement in aviation is likely to increase its own field, thru expanding the industry more than it cuts its own share.

But numerous inventions do not save anything, but give a new or improved product. Such inventions are bound to increase their own field (other things being equal)—(Prin. 10d), and seem to number about a quarter of all among the most socially important inventions of the last generation.[32]

The more narrowly an economizing invention is defined, the smaller is apt to be its share in a total process of production, tending to make it self-stifling; but on the other hand, the more apt it is to have substitutes, and thus be self-expanding.

But DO Inventions Go Stale?

While these considerations seem to forbid any uncomplicated generalizations about the effect of inventions upon their own importance, there remains Kuznets' list of large inventions with diminishing rates of improvement, asking further attention.

First one might cite counter evidence, such as the steady progress shown in numerous measures of the automatic stoker,[33] and the graphs of increasing steam pressure and coal economy in Figure 41 of our companion volume. When these last are plotted on semi-logarithmic paper they become nearly straight lines, indicating progress at rather constant rates.[34]

The impression of decline may be illusory for the reason that all writers, and especially social scientists, get their impressions of invention from books, not at first hand, and the books are always more or less behind the times, perhaps ten years, or fifty, being made in turn chiefly from other books. Hence unless the writer is particularly interested in present-day technology he will know little modern progress to record, and will probably give us another disquisition on the textile inventions of the Industrial Revolution. But even to the well in-

[34] Coal consumption is seen to decline quite steadily at 18 per cent per decade, and steam pressure to rise at 105 per cent till 1870, and 40 per cent since then, per decade.

formed it takes time, commonly half a century, for an invention to reveal and prove its great significance.

Why they do often Decay

But accepting the fact that very many inventions have declined, our first explanation may be that none could rise in relative importance unless others declined. As civilization grows more complex, as more things are added unto us, we might expect each invention's relative importance after a while to diminish.

Perhaps the greatest reason why inventions seem to decline, is that they change out of their definition. Tarde half perceived this when he said, "An imperfect machine survives, by a sort of metempsychosis in the more perfect and complex one which was in whole or in part the cause of its annihilation."[35] The steamer declines, according to one definition: but we might define it otherwise and say that it is changing into the motorship, and fast progressing in average tonnage and design. If an invention were defined with highest precision, it could never be improved at all; for the least change would move the thing out of the definition. We are back again to the perception that the definition makes the invention, which probably has no definition aside from language and other social habits.

Be the reason linguistic or inherent, it is certain that most inventions decline in the course of time, and that many may be said to die, like the watersail and the horse-boat, but equally certain that few die altogether. (Principle 11). Civilization is cumulative, as Ogburn has strest.[36] And as we said in connexion with the prophetic

---

[36] *Social Change* (N 169), pp. 73-9 on "The Cumulative Nature of Material Culture and its Diversification."

Dr Sanders reminds me that the making of inventions depends not merely on familiarity with the needed elements, but on recognition of them *as*

boats, cumulative, expanding civilization creates such diverse nooks and crannies of peculiar need or tradition, that almost anything can survive somewhere. Even the dugout, watersail, Roman topsail, ram, bonnet, quarter rudder and painted sail still have a few good uses among the advanced nations of the world. Numerous others have gone totally out of use. Even these are still alive in the memories of some men, or in some documents, and hence available as hints for further invention. But the change of requirements from age to age makes discarded arts rarely useful again in the same form, as we said anent Prin. 12. The only instances we can find in marine history of the revival of an invention which had apparently past totally out of use were the second and fourth masts, topsail, and the Renaissance revival of the merely projected Byzantine ox-paddle-wheel galley.

Some have thot there has been a general decline of man's inventive output; as we can by no means agree. Their reasons have been either the aforementioned insufficient acquaintance with recent and nascent inventions, or the similarly based idea that most of the possible discoveries have already been made,[37] or observation of the very real tendency toward decline in statistics of patents. We shall discuss this later in connexion with the patent

elements, which might be combined in new ways. We naturally hold our ideas in groups, configurations, *Gestalten,* with fixed internal relations which we can hardly think of varying, since we think of the configuration rather than of the elements. For instance, everyone was familiar with a boat's hull, and with water, and assumed necessary a fixed relation between the two for navigation, viz. that the hull should float on the water. That the hull should be lifted into the air, and set on legs on the water, was too absurd to imagine, tho I have written how it has been profitably done in the ladder-boat. The idea doubtless became imaginable when the invention of the hydroplane, which rides on instead of floating in the water, had shown men that flotation was not a necessary relation between hull and water. (To be sure, there were airplane and submarine suggestions too.) We take for granted that a boat must be symmetrical right and left, and differ fore and aft. Perhaps a flying proa of the Caroline Islanders, which sails wonderfully thru reversing those traditions, will give someone a good idea some day. Cf. ftN *86.*

system, and find little reason for thinking that invention itself is slackening in the age-old acceleration which Hart has ably sought to measure.[38]

An Invention is a Wide-rooted Complex (Prin. 2)

The question "What is invention?" might be answered from various points of view. There is the legalistic aspect, about which libraries have been written, the engineering and chemical side, the historical, the psychological,[39] all with their treatises, and good ones. But there is not one book yet on the sociological aspect, unless we count here Rossman's excellent *Psychology of the Inventor*.[171] We have been trying in our latest pages to answer the question linguisticly and philosophicly and statistically, and so were led into considering the cycle of inventions, especially their declines. Let us now make a more sociological approach, as we ask once more "What is an invention?"

Our 2nd Principle defines an invention as a complex of elements much more numerous and diverse in origin than any patent lawyer would think of, when drawing up as long a list of patent claims as he can. The invention is made up not only of engineering and chemical ideas from the prior art and science,[40] but also from considerations of the price of each material, fuel etc. and occasionally its supply, the climate and other elements of the natural environment, consideration of the skills, shortcomings and educability of the workmen, the wage rate and supply of the different classes, the degrees of inhuman labor and health risk to which changing law and custom allow subjecting them, the existing productive equipment that must be directly or remotely employed, like machines, factories, fire-protection, dry docks, railway tracks, hardware shops. Then there are the tastes, knowledge and attitudes of the consumers, the possibilities of modifying them, the size of the market, grave financial problems of how the invention can be developed and exploited, and legal problems of whether the patent system is likely to protect or ruin the backers of such an invention. And there are many more questions basic to the invention—every question that an inventor and his patron and lawyer would take into account, and still others that they would overlook, which later would unexpectedly throttle, foster or divert the invention. Every one of these goes to make up the invention; if any element changes the whole is likely to change, flourish, die or be inhibited before the start. As MacIver says,[40] the environment is responded to as a unity, both the outer aspects of civilization, or the means of doing things, and the inner aspects or culture, the goals, valuations, tastes and distastes that lead men to act. Any factor may be considered singly, but no one, nor dozen, can be set down as *the* explanation of the invention. And in the present flimsy state of our science, we cannot measure the relative importance of the factors—never well statistically, and usually not even by an informal guess.

The social causes of invention all come from the world outside the inventor and act thru him, (Prin. 27), and must consist of *changes* in the outside world, except insofar as a time lag might be said to occur within the inventor, requiring him to wait for a lucky accident or train of thot, or the completion of a series of experiments, or even conceivably to grow up as an appropriate genius, before the proper combination of elements of the invention may be complete. Principles 23 and 26 declare accidental invention and unique genius to be very infrequent causes of delay or acceleration. And in our history of the ship we have so constantly noted causes for the inventions, in changed circumstances confronting the inventor, that we have propounded our Prin. 7b, that invention is commonly adaptation to external change, which, both enables and elicits further inventions, as De Leener emphasizes.[188]

All will agree with the statement in our PRINCIPLE 6a, that the total *milieu* of the inventor, the surrounding situation, is changed by the accumulated labors of the inventor himself and of his compeers, creating a new "prior art."[41] 6b is not so well realized—that the non-technic factors cited under Principle 2 change too, thanks to individual initiative, propaganda and general social forces, with powerful effects. For example, why has the age-old progress in size of the biggest ships been halted for the last 20 years and seems likely to definitely end with the *Normandie* and 1018′ *Queen Mary* now being unwillingly completed? The reasons are not alone engineering changes, such as I predicted[303] in 1912 to stop liner growth about 1935, but the World War, the growth of economic nationalism, and above all the restriction of immigration to the U. S., with all the remoter social causes of each. The heaviest and hardest block of steel is molded into its form by human interests and emotions. The inner culture sets the goals, for which the inventor contrives the appropriate tools of civilization, while also in turn doing his part to modify the culture, providing a material basis for changed ideals.[42]

There are many social forces beside ideals and inventions, which are of vital, perhaps even predominant weight in the evocation of further inventions. Vital must be the growths of some all pervasive characters of our civilization — the population, age distribution, wealth, size of the business organizations, urbanism and

---

[41] Tisell (N 221, pp. 23-36 etc.) gives numerous stat. examples of the rise and fall of German patents in various fields, e. g. gas lighting, in response to changed industrial needs, or to the opening of a new and attractive field by some discovery or pioneer invention like Welsbach's.

[42] But we must not make the mistake of thinking that all inventions are adaptations to social needs. They may be the reverse—contrived for the private benefit of an individual or firm *against* society, as we shall make clear in our later discussion of Prin. 37.

industrialism. On some of these points I have indeed secured a few weak statistics of correlation,[194] between the patent scores of the states per white males aged 20-44, showing [43] correlations of .76 with percentage of men engaged in manufacturing and mechanic industries, .84 with wealth, .64 with secondary school attendance by boys, .37 with collegiate and professional school attendance, and .69 with the most fundamental variable, *climate,* as affecting energy and health according to Huntington.

No correlation alone proves anything as to causation, and we might have found correlations about equally high with crime, or Catholicism. For all these variables are associated with *industrialism,* and this, Huntington has shown, is localized above all by good climate. An isoplethic (or "contour") map I have made, of American patent productivity per capita, shows a heavy concentration in the narrow belt of best climate, near the 50°F. isotherm, from Chicago to Philadelphia and Boston, the same isotherm which I have shown [291] to mark the modern crest of civilization the world around, and the very same belt that houses all our first-class factories which had free choice of location, with rapid declines as one recedes from the belt, except for minor rises on the Pacific Coast and Florida . . . . Tho these correlations prove nothing alone, they are very suggestive support in connexion with the other evidence for our Principles 15-18, 22, 28, 29 and 33.

Importance of the Quicker Changes (Prin. 7)

While all factors in the total situation confronting the inventor and his patrons have their influence, even if like the force of gravity they never change, those that

[43] Pearsonian correlation was used for collegiate education only, the other distributions being non-linear, and for 3 others Spearman's rank correlation, which gave .56 in the high school case. For wealth, with a non-linear

have recently changed quickly and notably have a special significance, in two ways. (a1) They may serve as what Chapin [45] calls the trigger or precipitating factor, whether or not they have reached any particular grade of development, so that the immediate consequence of their coming is greater than they really caused alone. (a2) They attract notice and stimulate effort to meet the new situation, more than do constant or slowly changing forces equally weighty otherwise. We should so understand Kuznets and other writers when they speak of inventions calling forth complementary devices, always illustrated by the classic textile inventions. We may say one "demands" another; but usually the only obvious and real factor is a rapid change in the price of some thing. The growth in size of ships did not demand the invention of a steering engine; it only encouraged it thru raising the cost of hand steering, lowering that of an auxiliary steam apparatus, and attracting attention to the possibilities of steam. On sail-ships the need was

distribution. I used a perhaps new variety of correlation. First a spot chart was made with the spots enlarged according to the population of the state whose two measures each spot represented; then the regression line was drawn free-hand, a good hyperbola, traversing in the middle the bent stream of weighted spots, as a (tested) line of best fit; then the patents Scatter of the spots about this line was measured, and the correlation computed thenceforth as usual. Primary horsepower used in industry correlated .5. Probable errors of sampling were ± .04 for Manufacturing ± .076 for Climate, and ± .08 for Secondary Education. Pearsonian r for Climate was .59, regression perhaps linear; and the data were obtained by taking the geometric mean of the score of each state in the maps of climatic energy and climatic health. (From Ellsworth Huntington: *Civilization and Climate,* 3d ed., Yale U. Press, 1924, 472 pp., $5; Figs. 34 and 36). Other data were from the *Stat. Abstract of the U. S.* for 1924, and the Census. On the questionable significance of patent stat. see Chap. II.

Prof. Lowell J. Carr kindly allows me to use some related correlations from his forthcoming book. Correlating patents to the states for single years, first with their percentage of population urban, he obtained an r of .83 ± .025; with primary horsepower, among 33 states, .49 ± .09; with per-cent aged 20 to 44, among 20 states, .80 ± .06. Correlating the total of patents issued by the various nations to the end of 1923, he obtained: with production of pig iron among 15 nations, r equals .72 ± .1; with publication of scientific journals among 28 nations, r equals .71 ± .07; and with automotive registration among 36 nations, .76 ± .05.

met by putting as many as eight men in a storm on the doubled hand wheel.

Rossman[216] has shown statistically how patents in military fields respond to the stimulus of war, and H. W. Dickinson[191] cites how the *Atlantique* disaster stimulated inventions to prevent fire at sea,[41,46] while Rev. C. S. Bullock's successful marine fire alarm was brought forth by the *General Slocum's* burning.

Only this Principle 7 can put a sensible meaning into the proverb "Necessity is the mother of invention." Under conditions of stability, the most necessitous countries, like Amazonia and India, are the least inventive; and Switzerland, whose needs might be considered as least, invents the most.[194] But when a need or opportunity has arisen quickly, as happens to all people but most often in the most advanced, kinetic civilizations, minds are most apt to perceive, and respond inventively.

CHAPTER III

FACTORS FOSTERING, RETARDING, AND LOCATING INVENTION

SPECIALIZATIONS CAUSING INVENTION (Prin. 17)

Spencer's great principle of *integration,* here the centralization of work for an ever wider area upon increasingly specialized producers, may not only be taken as the very measure of civilization, but as one of the greatest direct incitements to invention. We may distinguish four kinds of specialization of labor, helping in different ways. (a), the geographic division of labor, enables each region to develop its naturally most productive lines, and helps invention by permitting (b), the specialization between establishments, since both pile up in one plant a larger quantity of work of one kind, that fosters (c), the division of labor between workmen, each with their special skill and tools or machine. This subdivided labor is of course much more productive, and two openings for invention arise, in the simplification of this split-up problem, and in the call to adapt production to a new method and ever increasing scale of output. For individual firms the market expands in spurts rapid enough to stimulate adaptive reactions, especially where a competitive situation shows the enticing possibility of a great expansion of output, if only some inventions brought an advantage over competitors. (Prins. 6, 7, 18). Particularly in the field of labor-saving inventions, large scale operation is crucial, as we shall discuss later with Principle 22. The invention is usually of a machine costly to devise, develop and build, and the accumulated overhead charges on these outlays can be recouped only thru an intensive use of this equipment of unprecedented productivity.

Adam Smith in 1776 wrote that the subdivision of labor (c) stimulated invention thru concentrating the attention of the artizan upon a narrowed field. Today this is probably negligible as a cause of invention, since the researches of Z. C. Dickinson [261-4] and Rossman,[171] and our own statistics hereafter, show that workmen contribute very few inventions of importance. The (c) subdivision does strongly foster invention thru making a worker's reduced share in production so simple, say just pulling a lever time after time, that it is easy to invent a machine to do that instead. Here we have a good answer, to those who carp at the Machine Age for stultifying work—that the more monotonous and stupid the job, the nigher it has been brought to abolition.

But many dull jobs, like candling eggs, have had inventive progress suspended because they require *seeing* colors or something, and no machine could see. Comes now the photoelectric cell which *can* see, to candle eggs or do a lengthening multitude of things better than the most keen-eyed, skilful, faithful and tireless workman, and to act electrically on what it sees, instantly, at any distance, with any fitting manner and strength. On shipboard already it can watch for smoke, in stack or hold, and see thru fog by infra-red.

One present backward and anti-inventive tendency, amid the age-old integration process, is economic nationalism. Its defenders say that that the tariff has built American industry; but it would certainly be much truer to say that free trade has. With 48 states and certain territories, covering a vast area and highly different in their naturally easiest products, all united in one free-trading union, of vast purchasing power, we have been enabled to extend Integration, the regional, shop and personal subdivision of labor (i. e. large-scale production) farther than any other nation, and for this one reason preëminently, to use more inventions, and obtain a greater output of wealth, than any other nation. Free trade did it, along with natural resources etc. Yet most people instead of wishing to extend the geographic division of labor wider, with all its beneficent consequences, ask to shut out still more the imports that interest them, thereby shutting in an equal value of exports, and so strangling some more of the little international trade or integration left. Here is a great reason why invention on the freight ship has not advanced faster, as well as on every kind of goods that might be economically transported afar.

The mass production of internally free-trade America is one explanation of the queer fact that our so inventive people took per capita but a tenth as many foreign patents as the Swiss,[47] and held twelfth rank among the nations. Our industrial methods, adapted to free-trade, mass production, high productivity and therefore high wage of labor, and low interest rate from relatively abundant capital, call for costly labor-saving devices which would not be worth using in the countries of small plants, cheap labor and scarce capital; so it is not worth while for us to patent such extensively abroad.

Ancient Rome—and the mention of those words sounds a warning, that every writer ascribes the fall of Rome to whatever social trait he particularly disapproves today—Nonetheless we shall go on to suggest that the noteworthy failure of Roman civilization to advance much industrially was largely due to a failure to extend integration thru improved transportation. For certain reasons,[48] they did not develop the merchant ship to be a good sailor and seaworthy, they tolerated piracy, and above all they had no horseshoes and clung to the

---

[47] 10.5 per cent, comparing white populations taking patents in non-adjacent lands of foreign speech, as in my article (Note 194).

[48] A more basic reason for the deficiencies of both transport and manufacturing was I think the poverty of the Mediterranean area in timber, metals, fuel, summer winds, water-power, and navigable rivers.

most ancient horse *harness,* which as Lefebvre des Noëttes [49] has brilliantly proved, was unbelievably stupid, strangling the horse as soon as he pulled hard, so that the heaviest load a normal ancient wagon ever carried was 500 kg, by two horses or oxen; with modern harness one ass will draw as much. With such harness and no horseshoes, poor rivers and defective ships the Romans could not, even after prodigal expenditures for paved stone roads, transport goods easily to mills and towns. So economic integration was balked, and civilization with it. Horseshoes, the modern harness and the steerable wagon appeared in the 9th, 10th and 16th centuries respectively.

Our fourth type of specialization, (d), is that of the PROFESSIONAL INVENTOR, or at least an engineer, chemist or factory executive who can give a large part of his time to improving his company's methods, and have a laboratory, assistants and other special facilities for inventing. The invention laboratories of the great monopolistic weak-current-electric, General Electric and Eastman Kodak companies are particularly well known, and illustrious in accomplishment; but there were 1000 research laboratories in this country recently, having 30,000 research workers and spending $200,000,000 annually, says Holland.[268] One-third of that money was put up by 23 bureaus of the federal government (but little of this went directly for inventions proper), and 1/5 of it came from the three great electric companies. The older industries, like the textile, have been the last to recognize the value of research; but that industry has at last established a trade association for research, such as spent 15 millions annually. Growth of organized inventing had been very rapid before the Depression.[50]

[50] Free-lance inventors such as Hatfield (N 161, 197), Fessenden (N 90) and Jenkins, often poohpooh the laboratories, I think with more bias than reason. Cf. Milton Wright on C. F. Jenkins, in *Sci. Am.*, Aug. '27, one of a good series of interviews with inventors.

A great laboratory, such as the Bell one with over 3000 employees, can further subdivide the business of inventing among an extraordinary variety of specialists, such as critics, microscopists, free-hand draftsmen, librarians, translators. Fullest information on past failures is made available.[51] And between inventors proper it brings the boon of interstimulation. The chief of the Goodrich laboratory lists as one of its advantages, that it enables keeping other departments out of the inventing business.[267] A great German laboratory finds that "about 80% of the problems submitted to them as new and requiring investigation can be completely solved by reference to their card index of past researches from all over the world".[52]

Whether these laboratories, or mere outsiders and free-lance inventors, have been responsible for the great forward steps of modern times is a disputed point we shall touch on in our later consideration of Insider and Outsider Inventors. We think the evidence favors the laboratories, e g Grosvenor's count[53] of 7/55 of distinguished inventions in 1889-1900 coming from corporations, and 5/17 since then, with the laboratories very rapidly growing of late. It is not to be expected they would have played any large part in the inventions of former years; and it takes time for inventions to be recognized as great. In such concerns as the General Electric, less than 1% of the valuable inventions are made by employes not hired specifically to invent. Holland says[54] that less than 5% of the American patents which reach commercial development and industrial exploitation are the work of independent inventors, and

[51] The value of such education in the prior art, enabling the inventor to begin where other men left off instead of starting from scratch, is upheld by Stamp (N 173) and many, but condemned by Georges Claude and some other inventors, as either discouraging, or leading into old errors. The past failures, Claude says, "vous renseignent . . . en vous cassant les ailes!" (Qq. idées sur l'invention et la recherche scientifiques; Sorbonne address; in *Chimie et Industrie*, 9: 1009-22, May 1923.)

that "the research laboratory has become the prime mover for the machinery of civilization."

Production Brings Improvement (Prins. 15, 16, 18)

We have discust before some departures from Principle 15, where patents on a device declined at the same time its production was maintained or increasing, and sometimes a stasis of unchanging design was reached. Yet in general this Principle that production brings improvement must be upheld, as one of inescapable logic. In our later examination of marine patentees, we shall find them 2/3 insiders, which at least strongly suggests that the more men employed in an industry the more patents will be taken out thereon.

Confirmation of this and Principle 16, that of fountain regions, is afforded by geographic studies, such as Tisell's (Graph 2) and the author's[194] comparisons of nations, and the interstate correlations we found above between industrialism and patent productivity. Since industrialism is associated with invention, and patents are taken chiefly by insiders, in their own industry, it follows necessarily that production in an industry is associated with its improvement (Prin. 15). The causation is reciprocal, the industry calling out improvements and improvements usually building up the regional industry. But always there are other factors entering. For instance, our rather industrial Southern Piedmont region produces few patents. The explanations probably are that the cotton textile industry is unprogressive, as Kuznets [166] claims, and that cheap labor, second-hand machinery and second-grade work do not encourage invention. To be a fountain region, a dispersion center for new inventions, production must be not only large but leaderly, meeting and solving new problems, (Prin. 7) not old ones, for which a solution has been already found.

Perhaps this Principle of fountain regions we have sufficiently exemplified, in connexion with North European sail inventions, pre- and post-*Clermont* steamboats, the propeller and modern rudders.

Our 18th Principle, that the firms leading in production tend also to be the leaders in invention, follows naturally from Principles 15-17 and 22. For such firms are commonly in the most progressive region, they can exploit the invention most extensively, and best use all the advantages of integration, inc. specialized inventors in laboratories; and they can exploit the invention with least obstruction from the patent system, since within their own organization they can collect the equivalent of royalties without a bit of the salesmanship, bargaining, cautions and quarrels which beset sales between companies, especially when the object sold is so uncertain in nature and value, and so easily stolen, or taken by inadvertence, as a patent right. Particularly when a small firm or an isolated inventor attempts to extract a fitting sum from a great or monopolistic corporation, is a patent right precarious.

Two Obstacles in Size (Prins. 19, 20)

Partial and quite minor set-offs against the strong advantages for invention which we have found in the big concern and industry, are found in the related Principles 19 and 20. The first cites that the standardization and tendencies toward uniformity which naturally accompany large organizations, whether a corporation, NRA trade organization, or a wise and leaderly government like Hoover's Department of Commerce tends in a few cases against invention. The Principle has rarely affected the ship, because ships are not closely standardized; but laws for tonnage dues, being uniform, have often produced marked distortions of hull, such as the turret freighter, and a Chinese junk with deck narrower amidships. American invention of car couplers has almost ceased since the adoption in the '80s of the Janney type, by law and what is more important, in over two million cars that must remain interconnectable. Waldemar Kaempffert has well developed this Principle.[55]

PRINCIPLE 20 finds a minor conservative, anti-inventive tendency in the growth of fixed capital, from the increase of wealth of a corporation or a country, and in the use of more durable materials and sound engineering and artistic design. Since such products are acceptable for a longer period, they tend to prevent new building and hence discourage invention by Principle 15. Wooden, ugly and inefficient America is being largely replaced by an America of pleasing and long-lasting masonry and steel; and in machinery iron is being replaced by non-rusting metals, or very hard and well-chosen steel alloys, making longer life natural even if for other reasons it fail to occur. The lives of ships are being lengthened by the return of rust-resisting wrought iron and by progress in navigation, fog-thwarting and general safety. Three recent ways of *welding,* invented all about the same time for the same ends (Prin. 34), by thermit, electricity and the blow torch, have also probably a capital-preserving effect, both thru building integral, non-deteriorating structures, and thru making easy the repair of fractures formerly irreparable, and the building up of worn parts. But on the other hand the blow-torch much facilitates the scrapping of obsolete steel structures.

The preservation of equipment checks inventions not only in the object directly affected, but in any others that must be used with it, as a railway track, bridge or tunnel controls all the rolling stock, and the ships the docks, and the docks the ships[200,201] . . . Kuznets[166] points out that the industries based on old inventions have in general more invested capital to block progress. . . . Sir Josiah Stamp [173] essays a mathematical determination of the proper time to adopt an improvement, and concludes that if one country be less progressive than another, it will not fall farther and farther behind, but adopting the inventions of the other will have a constant lag of 10-15 years at most.

These Principles 19 & 20, account for much of the charges commonly brot, that great corporations wilfully and wickedly suppress patents. Most of the remaining cases can be accounted for by the fact that corporations take out or buy patents on alternative ways of doing the same thing, while outsiders, especially the author of a supprest invention, may well think erroneously that the means left unused was the best one; and finally by the Principle that inventiveness itself may inhibit the use of invention, if it lead the enterpriser to think that some still better means is just around the corner in the future, so that it were better to hold up expensive rebuilding until a nearer final and perfect plan be at hand.

Resembling fixed capital are the specialized skill of workmen, knowledge in the higher ranks, and the good-will and going-concern value of firms, painfully acquired thru years of work and advertizing, and subject to most destructive devaluation by all inventions that radically change things. Self-interest naturally inclines all the men menaced to fight such inventions. When labor unions do this they are severely lectured; when business heads do it it is usually taken for granted that they will act, or be inactive, in their own business interest.[56] The economist, with clear sight and the interest of all in view, will sympathize with both workman and employer, and seek means to mitigate the personal destructiveness of inventions which must come, by a real program for redi-

[56] But for Europe Jas. Douglas tells of sentimental ties uniting firms and employes for generations, which resist labor-saving machinery. N 264.1, *ca.* p. 36.

recting and retraining the displaced men. And he will recognize the destruction of mental capital as an actual loss going into the calculation of social desirability, so far as, and only so far as new learning, like new capital and construction, will be required, and other costs of transition, especially the unprevented suffering of the workmen, are incurred. There is no loss simply in destroying no longer needed capital—the loss is in having to create new. Nor is loss of commercial good-will, including dearly bought reputation from advertizing, a social loss—it is only a loss matched by gain, between individuals, the national wealth being unaffected.

Population Influences

The great social force, population growth, now apparently drawing to an end in America and Western Europe, has been a stimulant for invention, in two ways. First by Principle 21, when population is rapidly growing, new equipment can be added without scrapping old, thus partly getting around Principle 20, above. The like is true, and probably more important, when wealth grows, and industrialism, or some industry in particular. We have been able to add new autos without waiting for the old ones to wear out; but not to add many new Great Lakes ships.

Principle 22 states that "Increasing population andor industry stimulate invention, because they increase the absolute need for a device, and the number of potential finders, while the cost of finding it remains the same." There are more mouths to eat the invention, so to speak, and more eyes to find it (Prins. 15, 27) the difficulty of finding it is no more than before, the value when found is greater, and so, many more inventions are found. We shall speak in connexion with Prin. 27 of the much vaster increase in the population *effective* for invention, thru technic and other education.

This Principle 22 is that of surplus overhead, and stimulates invention in the same way that it drives railroads and other businesses whose outlays are chiefly for overhead, for equipment incompletely used, to seek further traffic with desperate avidity, because the added gross revenue would be mostly net profit. Invention and other kinds of innovation differ from all other pursuits of mankind, in that the proportion of overhead to operating cost is very many fold. That is to say, the devising of an innovation costs about the same, whether it be used much or little after discovery. The only operating cost, that is the only cost that increases with increasing population or use of the invention, is that of making the innovation known and used; and this work will carry itself if the innovation be popular. More population does not help with more portrait photographs, because more labor must go to make them. With food and all land-based industries, more popuation makes the situation worse. But as to invention, increasing population or industry entails that each inventor's work is more widely useful than before, at the same time that there are more inventors to work. So here are two reasons why increasing population fosters invention; and the two reasons are multiplied together, not added. This double Principle,[57] with its extension to cover technic education, is one of the greatest reasons for the increasing rapidity of man's technologic advance, which Hart [159] has sought to measure. Moreover, from its simple, mathematical nature it can be affirmed with certainty, altho we do not know how to measure its effect in practice.

[57] It was pointed out by Condorcet in 1794, in his *Esquisse d'un tableau historique des progrès de l'esprit humain,* while Turgot before him saw that the more geniuses the more progress in science.

Increases of population and/or industry entail of course many other effects (notably thru the land/man ratio), which any economist can predict as readily as the author, thru our Prins. 2, 6, 15-18, 21 & 27. Yet all the unlike effects are probably small compared with the consequence described above.

Invention costs run to big money. Any radically new design for a gasoline engine will take at least $80,000 before the first one is operating satisfactorily.[275] A new typewriter's costs are higher; a new kind of large ship will roll up the millions. The increase of population helps to justify such large initial outlays; increase of business does more; and increase in the prospective business of a particular big firm does most of all. (Prin. 18).

THE SCOPE OF CHANCE (Prins. 23b, 24 & 14)

Our 23d Principle declares that inventors are led to their task by perceptions of its need and possibility, and that invention by accident is very rare or unknown in marine history. Such invention is frequently talked about, and has been very soundly explained by Dr. Rossman [58] and Mach.[167] But it is perhaps not so common as thought, since from all our studies in the history of the ship we can recall but a single case of it. Marine equipment, contrasting with chemical, electrical and optical, usually needs to be so large and/or in dangerous situations, that it is not handled in the haphazard manner most productive of lucky accidents. Our one case was a typical accidental invention. F. P. Smith was propelling an experimental boat by a bad, long, Archimedean screw; a sunken obstruction knocked off part of the screw, whereupon Smith noticed that the boat went faster than before; and so he improved his screw accordingly. Notice that the accident happened to a devoted inventor, who was trying to find just that which the accident taught him. As Dr. Rossman says,[58] "In nearly all cases we find that the accident happens only after a persistent and carefully conducted search for what is wanted." Some percentage, perhaps small, of lucky accidents is a normal feature of intelligent and well-implemented effort to invent, usually in that same line, and always occurs to a deserving person. If the lucky accident happened to an ordinary person, it would not be lucky, but merely another accident, suggesting nothing, or at least nothing that would be laboriously carried thru.

We must now recall again our consideration (under Stagnation & Prog.) of how the time distribution of inventions along the history of the ship appeared the more regular and explicable, the more the number of inventions considered was enlarged, whether by considering less important inventions, or by enlarging the definition of the kind of ship or number of patent classes or countries considered. But now our 24TH PRINCIPLE denies that there is any chance at all, regarding the latest date at which an invention can appear—that be born it must when its time is ripe. Let us explain the apparent inconsistency.

INVENTIONS CAN BE PREDICTED (Prin. 14)

Each invention, we have insisted, comes just about when social forces bring it (Prin. 24), thru the entry of new forces as per Prin. 7. And we can largely perceive and understand these forces, as we saw why steam steering gear, or 2000-ton windjammers, had to come just about the time they did. We may even understand the forces ahead of time, as for instance engineers predicted the 1000x110′ ship limits for which the Panama Canal locks were designed in 1906, or foresee the present end of the great liners, as the writer did in 1913.[303] With less knowledge, we may say that the Occidental rudder probably had to originate about 1200. With much less knowledge, we may guess that the fore-and-aft rig had to start about 800, for some reason or other—or that it did not have to, because of any social, economic forces, but only because of some little contingencies that we may as well call pure accidents, such as the birth of a genius, or one of the chains of events which we suggested, happening to take place for any reason. The lateen or the lug sail is a simple thing, and apparently would have been possible and useful for the previous 5000 years. We may say, then, that some inventions like the lateen and all later new sails, and all inventions before the Renaissance, do seem to have in their date of birth a large element of irreducible chance. But the greater our knowledge of the historical situation the less, usually, seems the field of chance. And today, with inventors far more numerous and active, the chance element is much less. The rotorship and the stream-lined rudder seem inevitable within a few years of their time; perhaps they might have come 5 years later, or 20 earlier.

But *why* inevitable? Because of a special, novel assortment of numerous forces and given facts in an ever changing world situation. We might understand all of these enough so as to predict single inventions positively, and yet there be no such rule as that whereas great fruitful contingencies numbered, say, 7 in the 1800s, they must number 16 in the 1900s, and not 13 nor 22. Chance means forces which we cannot understand, either because of their triviality (e. g. depending on one personality), or from the undevelopment of our science. Insofar, then, as inventions are determined by chance according to this definition, their occurrence should be called sporadic or haphazard. They might be such in the historical graph of frequency, altho each invention individually was definitely timed, understandable and predictable. Then we should remember that the irregularity of their plotted historigram will be reduced and rationalized, the more we increase the number of inventions considered, as in Graph 3.

We have further discust the prediction of inventions under "Ships of the Future" in Chap. VI of our companion volume. Understanding the forces at work is not the only way to make predictions. As Usher says,[59] when inventions are viewed *ex post facto* they fall into orderly sequences. Even with no understanding we may perceive a trend, and project that into the future. To be sure, the probable error increases fast the farther we wander into the future. With no idea of causes we might with considerable security predict further travel by the ship in her established trends of greater average size, with length increasing faster than beam or depth, greater speed, power, safety, reliability, comfort, economy, labor-saving, strength, lightness, hardness, and beauty.

This apparent tendency of industrial evolutions to follow definite lines is strikingly similar to that observed in biologic evolution: what the connexion, if any, I know not. If we may indulge in a little schematization, let us propose four historical stages for it. First, in Savagery, the changes and their direction were unconscious. Then in some cases men perceived that their own ships were different from those of other nations, by reason of certain pervasive traits like smoothness of line, and set out nationalisticly to accentuate or at least preserve these characters. The peculiar and extremely long-lasting Egyptian hull form might be so explained. A 3d or often 2d stage would come when with a little historical knowledge people would observe that progress had occurred along certain lines, such as increase of masts, and so be more inclined to try further steps in that than in some strange direction, such as changing to an A-mast. The last stage is of the corporation and its inventors, fully conscious of history, trends and the future, and with a long-range development program.

The simple extrapolation of observed trend, such as we have been speaking of in the last two paragraphs, is much safer when combined with the rational process of understanding causation and considering the forces, as previously. Thus, ideally, we determine the trend, with direction and rate, for each separate force, decide by reason how the forces will interact and which will prevail in each future weight-combination, and hence plot the joint effect upon our ship or what not. Thus one might avoid the mistake of other prophets of future biggest liners, who paid insufficient attention to causation, and hence predicted constant growth, instead of the imminent decline in maximum size, which will come thru air route competition and the other reasons I gave.[303]

Searching for Abrupt Inventions (Prins. 1, 2 & 26)

As another test of the certainty of timing of inventions, and of the generally believed necessity of the particular inventor, or of genius (Prin. 26), and of Principle 1 of Evolution, let us sail again over the course of the ship, seeking only this:—important sudden inventions, that were *not* evolutionary, but strongly inventive, winning by one stroke what was *not* an obvious nor clearly inevitable development, but was a technologically revolutionary step, founding a new line of mechanisms. And to have a limit to our list,[12] let us stick close to the marine inventions in the lines of direct descent between log and rotorship, as we intended to do, but have often departed somewhat from, seeking enough inventions to discuss. In the premodern history of the ship we found evident a slow evolution of some features; lack of data obliged us to propose hypothetic evolutions of some others; but nowhere did we find an invention which seems likely to have required a decided contriving, somewhat ingenious by modern standards—except *perhaps* in clinker building, the marine hodometer and its offspring, the ox-driven paddle-wheel, reef-points, the lateen sail, rudder, jib and stuns'l. The other inventions, like the fore and mizzen masts, topsail, etc., seem too simple, easy and inevitable to deserve place in our list of perhaps revolutionary inventions. For this list contemplates possibilities of the invention failing to be thot up; but such inventions seemed to arise inevitably when the size of ships, or other circumstances, called for them. To be sure, they might have been rather difficult and delayed, in an age so uninventive that any altered disposition was almost unthinkable.

When we reach the days of steam and oil power, important abrupt marine inventions are faster coming, but fewer as yet in total. We might list the propeller, but Leonardo's air-screw, the first one, was not a marine invention. Nor was Flettner's current-actuated rudder. The contrapropeller and the waterjet are not used in the rotorship or its ancestors. Engines and metallurgy were developed first ashore, water-tight bulkheads were used in Hindu ships of the 1200s. The chronometer is hardly part of the ship, and had a gradual and duplicated origin anyway.[61] The rotorship should probably be counted, tho inevitable. So we find only a couple or so of abrupt inventions at the most, in modern times.

If we went over the foregoing list of a dozen, or some such little number, of distinctly original, abrupt and important inventions made in connexion with marine ancestors of the rotorship, we should be further struck with the fact that not one of these ever revolutionized the ship, nor was itself entirely without precedent or starting points. Several, like the hodometer, lateen, stuns'ls and chronometer, might well be called evolutionary. Neither as to causation nor as to results do we find much of revolutionary inventing.[62]

And yet, with only a possible dozen of important abrupt inventions made on her, our ship has somehow developed all the way from the log, most fit for amusing a swimmer, to the rotorship *Barbara,* that speeds 2830 deadweight tons across the seas, safely and swiftly, by lofty towers spinning the invisible gale.

A New Kind of Invention (Prins. 1, 31)

Shall we leave, then, this list of a dozen or so of possible abrupt inventions, as the exceptions to our general rule of evolutionism? They might be justified as such by the law of chance, which probably should leave a small proportion of the cases as extreme, atypical examples, some of extremely gradual development, some of the opposite—rather abrupt appearance.[63] However, we think that most or all of the more abrupt modern cases can be explained by a consideration of the methods of modern invention. It is chiefly the modern which exemplifies our point, because there has been a trend for ages (Prin. 31), away from the almost unconscious, mere evolutionism presumed for primitive cultures, thru the more

---

[63] Since the cases (inventions) reflect the operation of many different causes and items, one would expect the frequency of the different degrees of evolutionary character to be plot-able by one of the bell-shaped curves so constantly recurring, tho not necessarily the normal probability curve. This is indeed assuming an essential unity, like that of biologic, sociologic and astronomic organisms, in the phenomena under discussion, here the making of nautical inventions. I think that such is supplied by the fact that a biologic and sociologic unit, the minds of English-speaking people, defines and delimits the cases to fall under "the making of a nautical invention," defines them by its habits in using those words. In any case this discussion should clarify our idea, to one familiar with Statistics. Assuming then as justified some sort of bell-shaped curve, most of the cases (inventions) that occur are more or less closely grouped about a central mode, or standard (m in Graph 4); but there occur toward each extremity a few cases, of rapidly decreasing number, of each degree of abnormality, however great, if the total numbers of cases and of factors be large enough.

Thus, there should be, among nautical inventions, a few of extremely evolutionary character represented by cases far to the right along the OX axis, and a few at the opposite extreme (left) that were more and more abrupt, in number of cases rapidly shrinking toward O. Suppose the cases at X_e and to the left of it depart so widely from our standard (the modal duration, OX_m) that they should be called exceptions to our 1st principle. Suppose now we find, as we shall next try to demonstrate, a single factor which explains the brevity of gestation of a fraction of all the cases (which are resumed under the smaller bell-shaped curve, shifted toward the left, the quick end of the distribution curve of the total). Now the degree of abruptness X_e has left unexplained only the small difference between the numbers of cases of X_eE and X_eJ, and the number of still abrupter inventions unaccounted for is still more reduced, from the triangle OX_eE to the triangle OJE. From the total area there remain now very few cases unexplained that are of particularly short gestative period.

revolutionary big-ship-building of the 1400s, to the deliberate, scientific research of laboratories today, and farthest of all to the logical, revolutionary inventing of a Flettner, who set out to alter the ancient ship fundamentally, and in his rotorship did it.

It looks somewhat as if our evolutionary principle of invention were being gradually superseded by one of wilful, abrupt creation of new things by inventors. I think rather that evolution is being switched in the scene of its action,—hid behind brick walls. It used to take place thru the embodiment of an idea in a public, physical ship, then the copying of that ship by the same or another designer, with a slight change introduced. Now, with the development of sciences, and a technique of ship designing and even inventing, the first small modification of the existent ship may take place only in the inventor's brain, to be followed by another and another in the same brain well trained to think and remember along these lines. But shortly even this powerful mind would become confused, so a sketch is made, materializing a part of a ship on paper. More sketches and complications follow, and presently an elaborate and exact mechanical drawing is made to embody this new evolving ship, and to test practically many of its proposals.[64] Next a model may materialize it. This further assists the inventor's mind to take other steps onward, from ideas that are new, yet firmly grasped, because they can be looked at and handled. Also the model, more than the most exact drawings, permits actual experimentation, to determine what the science, the "method inventions" were not powerful enough to predict. Models have been especially important in the recent development of the ship's underwater lines, propellers and rudder, thru experiment with one form after another, towed or driven in a tank. For a fluid acts somewhat the same with large and small sizes, more than machinery, and the differences, due to scale effects and the nearness of the tank walls, can be pretty nearly calculated by modern hydrodynamics. The last embodiment of all is in a full-size experimental ship, such as Smith's *Archimedes,* Flettner's *Frigido* and *Buckau,* or the *Carmania* which the Cunard line built identical with the *Caronia,* save for giving her their first turbine engines, with triple screws.

When all these successive ideas and embodiments of the new invention are made by one chief inventor and a shipbuilding group, who can set their minds to a problem and great material means too, so as to follow it all the way thru, most efficiently and with no costly errors, a strikingly novel invention will sometimes reach the stage of commercial practice before the public hears anything about it. In this case it seems to the public that an invention has been made overnight out of nothing, or out of something quite different; but the long-laboring designers know differently. It is evolution as of old, but far quicker and more efficient, and out of public sight.

The inventor Flettner writes [65] "The reader . . . . will learn that the development culminating in the ship's rotor took place systematicly step by step and that each invention always contained within itself the idea upon which the foundation of the subsequent invention was built; like links in a chain one idea was tied to the other, and all my inventions after all are based on one common principle. [The application of flow-laws to useful purposes, or perhaps one can say, of Bernouilli's famous equation, or the fact that a useful pressure may be obtained across a stream, by producing accelerated and retarded flows on the two sides of a solid body.]

There has developed another change in invention. Bernard [66] calls the modern inventing "projected," since it is largely thought out first or drawn or modeled, and the old method "empiric" since it proceeded rather by practical, full-scale trial and error, much oftener error than improvement. The "projecting" mind, particularly of an inventive genius, may make great leaps from one field to another, carrying a bare idea with him. Instead of having to derive an improved mast simply from an old mast, as always before the 1800s, he can derive it from anything—from an iron pipe or a wicker basket, a tripod or a latticed tower, to get certain masts of today. In each case his mind seeks anywhere, to borrow for the basis of his invention, embodiments of some few pure principles, that he sees are necessary or desirable in a mast—such principles as that any mast is a beam, needing strength against bending, most at the deck level, thru resistance to deformation, etc., and that it needs slenderness to the wind, and light weight. A wicker basket with crost diagonal stakes offers those traits; therefore the ingenious engineer takes it over to make a "basket mast" on his American warship. From the marine point of view the new mast looks like an utter invention. Nor would a willow basket maker call it the latest offspring in his ancient line of wickerwork. But from the engineering point of view it is nothing very new—it is an old principle used for a new purpose, like a hand shovel giving principles to a steam-shovel, or a mast to a flag-pole. So a Flettner's mind passes from steering an airplane to steering its rudder,[67] then from using this current-actuated rudder on an airplane, to a ship: again from the across-wind push of a cylinder revolving in a wind tunnel,[68] to a similar cylinder revolving in the ocean wind and pushing a vessel. As a ship the resultant rotorship is the most unprecedented craft that ever floated; but the mechanical principle of her motive power is 50 years old, and had its begin-

[67] One would gather from his book that his route was thru aerodynamic experiments with rudders of various degrees of balancing.

[68] Plus the idea of a revolving skin, as narrated elsewhere.

nings in knowledge before that. Usually a principle is more modified and disguised, as in the basket mast, when carried over to a quite different field; so Bernard calls the modern, telic, projective inventing "indirect or abstract copying." Copying and a sort of evolution it still is, but the thing copied is thinner, more abstract.

The modern marine engineer is acquainted with many principles, many inventions of the past, many products of Nature; and if he has the inventive mind it will persistently soar like an eagle over many fields, swooping to snatch a useful principle from anywhere, so that the primitive or ancient inventor compared to him is as the field-mouse to the eagle—purblind and busy in his little runways, altho his brain may be as good as the eagle's, and his thought processes the same (Prins. 11, 5).

Yet even the greatest genius can change little about the thousand-sided ship, as a glance at any consecutive pair of our pictures of ships would show.

CHAPTER IV

THE INVENTORS OF SHIPS AND THINGS

THE UNIMPORTANCE OF THE INDIVIDUAL (Prin. 26)

Only in the dozen-odd cases that we found, of important inventions perhaps made abruptly, by a large and difficult leap forward, can we imagine that the death in infancy of some inventor genius might have changed or delayed some serious feature of the ship. And even with this dozen, further knowledge of the mechanical surroundings, the industrial situation, and the events at the birth of an invention, such as full data enabled us to make for the rotorship, would likely convince us that it was either inevitable then, or unimportant then, altho it became important centuries later, or that there were easy substitutes for it, like the wood-and-copper sheathing that substituted for anti-fouling paint.[69]

"For let Philosopher and Doctor preach
Of what they will, and what they will not—each
Is but one Link in an eternal Chain
That none can slip, nor break, nor over-reach."[70]

[69] Kroeber has made an interesting statistical demonstration of the unimportance of the individual in a field which might be peculiarly subject to his influence, that of dress fashion. He has plotted regular curves, with suggestions of cycles, in the changing dimensions of parts of women's evening dress from 1844 to 1919. He points out that the common man is interested in his own feelings, morality, and fate, so reads the same into history. "The satisfaction of these interests has its justifiable function; only it prevents instead of cultivating an understanding of the workings of civilization. The individualistic view of historical phenomena is in its nature subjective, and its treatment must always remain subjective." On the Principle of Order in Civilization as Exemplified by Changes of Fashion; in *Am. Anthrop.* 21: 235-63, esp. 262.

Yet there are Captains of Progress

We have, however, often hailed geniuses on our voyage, especially of late. An apparent entry of personality has occurred in the last two centuries, along with a profound, it seems an utter change in the ways that inventions and other marine betterments come about. To be sure, the apparent absence of genius in earlier centuries is likely largely due to the oblivion of the centuries.

> "Many a hero now forgotten sleepeth quiet under
> ground,
> And upon the earth no echoes of his glory ever
> sound." [71]

For it is only of late that some people, hardly the Historians, have taken to writing down the principal facts of man's technic progress. Some myths there were. But whatever the single services of individuals, we may doubt that there were ever great and successful marine inventors before the Renaissance, when there was no profession of inventor, nor usually of engineer, nor any way to make a living simply by improving the design of ships or anything.

What the Great Do

But if genius is certainly frequent latterly, must we not infer, in spite of all evidence we have adduced from the inventions themselves, that great personalities have after all been latterly necessary to the development of the ship?

Our answer must be both yes and no, if we examine the activities of those great men. First we found an illustrious series of great mathematicians and engineers inventing screw propellers, a century or more before they were wanted or conditions were ripe for them. Then in steamboat inventing we found numerous men of ingenuity, some with egregious pertinacity, and just before commercial success three men, Evans, John Stevens and Fulton, who combined every gift and acquisition of genius, beside Symington who was not so distinguished, but made a perhaps better boat than any of them. There were numerous other geniuses in the ship's history—R. L. Stevens, Ericsson, Smith, I. K. Brunel, McKay, Scott Russell, Froude, Diesel and Flettner, who are certainly divine men, to reverence and bless for the great work they did. But what was their work? Some great inventions that they originated and brought to large utility? Who can name one, that really knows his nautical history? We repeated a long list of R. L. Stevens' excellent steamboat inventions; but we search it in vain for a critical discovery. Scott Russell started the longitudinal system of construction, but it was not developed nor used much for half a century. Ericsson's *Monitor* was an important invention with unusually little precedent, but it does not fall within the elected peaceful field of our inquiry. McKay built wonderful clippers, but never made an invention that we hear of, unless in adding a fourth mast to his unprecedentedly huge *Great Republic,* and two topsails in place of one. And so for each other genius in the wake of our ship—if the reader will but refer back to his work, he will find it by no means indispensable. What each did was clearly either inevitable, or not then valuable.

But all our explanations of individual non-necessity have not abolished the great men. There they stick, like volcanic peaks above ocean and cloud, whose Olympian gaze confronts us long after we sail by. Can we call but a mirage the stone and fire of the Peak of Tenerife, or soon drop it below the horizon?

No; but this we may say: that the many, many services rendered by each genius show by their very number that they were not of a unique and indispensable order, but such as are commonly rendered by a number of smaller men together—or by one genius instead of another. We cannot find a single example to offer, but if there were some desired discovery so difficult, so recondite that only five men in the world had the intellect and the preoccupations which would lead them to it—not one of those geniuses would be necessary for the invention, since any of the remaining four would find it instead, if one were removed. Economics teaches us that where there is no scarcity there is no value nor price, even for an article as indispensable as air. And there is no scarcity where the supply is greater than all the demand. For inventions or any other kind of innovating it is true enough for momentary purposes to say that the whole world's *demand* is for but one man. *One* can make the invention, and then all the world use it. (Prin. 22, previously). The modern world *supply* of available geniuses is sure to be greater than one, even in our extreme supposition that only a genius can make this invention. This is assured, as Prof. Ogburn has pointed out,[72] by the normal probability curve which governs the distribution of brains and every other quality known, and is confirmed by the study of inventions, especially his of independent duplication of invention. (Prin. 25). In the electrical field this has caused rival claimants for every important invention. And Rossman points out,[73] that about half of all inventions that pass the stage of patent application are thereafter dropt because of the discovery of prior inventors, not to mention the number dropt for this reason at earlier stages. In the stricter, German patent office, only 37% of shipbuilding patents applied for in 1877-1907 were granted.[221] Inventors are constantly advised to keep proofs of their priority, their dates of conception. Bernhard Stern has well demonstrated the abundance of duplicate discovery in medicine.[74]

Now to bring our hypotheses on Prin. 26 down to practical earth. On the one hand, modern history gives no reason to think that any invention would be worth making, (Prin. 36) in an age when it was so difficult and recondite, so far from contemporary practice, that only five men in the world could be expected to make it. A scientific discovery might be worth making then, but not an invention, which must be generally welcomed, to be of much use. On the other hand there are not one, but millions of practical problems to be solved, a number far greater than the total of potential inventors. So each good man must make more than one invention; the best glory of the inventive genius is that he makes hundreds, as Carr's,[185,6] and Rossman's [171] patent statistics show; and still most treasures must be left unfound. So the demand for inventions *in general* is greater than the easy supply, and a high value attaches to inventors, especially great, prolific ones. However, this shortage does not permit highly important problems to go long unsolved; for clever brains are ever roving over their respective fields, seeking the most important and remunerative problems to solve next; and such a gem that mankind's diggings have brought to the surface will not go long unfound by someone.

For example, we can not merely assert with confidence that had not model-towing experiments been made by the genius Froude they would have been made by someone else, genius or not, and the same improvements of ship form found out. We can almost prove it by the frequency of patent interference, or by the fact that not only Froude, but Beaufoy, Condorcet, d'Alembert and lesser men from 1767 onward had been making model-towing experiments. Or we may recall the many minds, great and smaller, that on the early steamboat and propeller were doing each other's work, especially Stevens and Fulton, two geniuses side by side, and Smith and Ericsson, accomplishing the same work in the same waters at the same time. And there are the lists of duplicated inventions, such as the Ogburn-Thomas one,[72] easily expansible.

But the commonest is for one man to carry farther, into practical use perhaps, what many others had handled in earlier stages. Our geniuses may be found in any stage: if one did not work chiefly in the commercial stages he could rarely earn a living; if he did not also pioneer, he would hardly be remembered as an inventor of the ship. The normal relation between inventors is sequence rather than parallelism and duplication, merely because no one will waste his time inventing the already found, if he can learn of it; and those with the brains and facilities for abundant inventing are well informed. The rotorship can henceforth never be reinvented, from that day when the *Buckau* "set sail" and the depths of every ocean tingled with the cabled news. But how many will live who *could* have invented her is indicated by the duplications of inventions less suddenly and universally made known. If duplicate inventing is less characteristic [75] of nautical than of other (esp. electric) branches, the fact (if such) is likely due to the publicity of ships, from their large size and traveling, and the irresponsibility of their crew. Often an invention, especially a chemical or electrical one, might easily have been made previously and independently, but never tested in practice, with the result that we never hear of the duplication.

In fine, we conclude that our age-old vessel has felt the tread of many great inventors, and that these men were, like the great in other lines, men of energy, sound emotions, impressive character, tenacity, vision, originality

[75] We have noted in our nautical study various cases of apparent independent invention, of the steamboat, propeller, and its increasing pitch and shrouded form, balanced rudder, tubular boiler, and steeple engine. But even between Orient and Occident it is very difficult to demonstrate independence of origins, and much easier to note rival claims.

and a general ability to create and do, far beyond that of ordinary mortals. They commonly use their gifts first to acquire a good tho not the completest knowledge of the business they will invent in, or one akin to it. Then as the natural guerdon of their grace and well-planned efforts, they are rewarded by inventing a million-fold more than the ordinary man, and a hundred-fold more than he ever *could.* Such quantity would be measured both by number of separable forward steps, and by the originality or profundity of a few of them. But there is no proof that genius ever made an immediately highly useful marine invention that was too deep for mere talent soon to attain. And if there were such a problem, we may well believe that other geniuses today would attack and solve it, if solvable, as soon as it became pressing and an evident barrier to progress.

The Mythology of Great Inventors (Prin. 30)

The popular belief in individual, single, great inventors for things has been grotesquely developed by the same process as that which built all the classic mythologies to account for the origins of this and that, so that our traditional great inventors, even when really great and the doers of much of what is ascribed to them, are still, from another aspect, mythic heroes in our school-propagated national epos. But we have probably discust this sufficiently in *Inventing the Ship,* anent the titular inventors of the steamboat and screw propeller, so need here add only one observation:

The mythology of heroes or demigods is built up, 1, by the interest of any group in having an heroic origin or history, so that the members may feel flattered by psychologicly identifying themselves with something great, and, 2, by their interest in having a common cult and symbols to increase the cohesion of the group and its loyalty to its living leaders, all of which is very helpful to these last. Thus each monastic order promptly has its Founder made a Saint.

Clans have gone, and among the strongest social organizations today are the corporations. To some of these the mythology of a certain invention is all that we have said, and a most valuable business asset, a well advertized sort of trade-mark, medal and reputation for being "the original." The American popular belief that McCormick invented the reaper may be worth millions to the International Harvester Co., the fame of Bell certainly is to the telephone company, Fulton is the hero of the Hudson River Day Line, Edison and Marconi of various companies, and the Wright Brothers of the Wright Aero Corporation, while the rival Curtiss interests set dead Langley on an airplane, like the Cid on his horse again, to fight their battle. And many a minor corporation industriously and probably with conviction gilds the sepulcher of its inventor-founder. In our age of propaganda (inc. education and advertizing) those ideas are likely to prevail which have the most deeply interested backers. Truth is a frail, consumptive lady, alongside Propaganda.

The reality of a great invention, as we have shown anent the steamboat origin, Prins. 1 & 2, and often, is a conglomeration of detail inventions so vastly numerous that no one hero, nor country, nor century is enough to produce it,[305] nor all the dictionaries enough to define it.

Inventors are the Medium for Every Force (Prin. 27)

However little necessity we have found in the individual inventor, we must avow at once that without some kind of inventor there can be no inventions at all. Moreover the whole class of inventors occupies a crucial position in the machinery of progress, since every fact that bears upon invention, whether to promote, hinder or direct it, must act upon and thru inventors. A social need can no more create its solution save thru inventors, than a good climate can invent. Hence we claimed that the inventors (as a class, not as irreplaceable individuals, and meaning all who invent, not merely the less important group of professional inventors) determine the directions, frequency and efficiency of devising, "in proportion to their absolute numbers, intelligence, moral traits, strength of motives for inventing, time free for it, and mental and mechanical equipment for it."

A strikingly rapid alteration in one of these traits, mental equipment, has occurred thru technologic and general education, so that the number of minds in America capable of making inventions has increased far faster than our census population. The collegiate students of Engineering in the U. S. increased about 90 times faster than population between 1830 and 1930,[75] making the absolute increase 940 fold. Those students numbered at the 20-year intervals approximately 50, 190, 400, 1195, 23,241 and 47,000 respectively.

Numerous and revelatory studies have been made of the personal traits of inventors and of their psychologic methods, some of which we have listed in our special bibliography for this field, q. v. We may stop to note here but a few of their findings. The studies of Rossman[171,215] and Carr[185,6] are specially notable for valuable statistics intelligently handled, from returns of questionnaires to inventors and from patent data. Both for instance reveal a strong tendency for inventors to come from families in the professions (4.5 or 10.4 times[186,171] the normal quota), and Rossman shows a decided hereditary influence in inventiveness. Carr finds that those inventors beget the most children who have the most brain-children in the form of patents. "Holding the different in ventor types constant, the size of the inventor's family correlates .878 ± .014 with the intensity" of the patenting rate. This reminds us of the important fact Huntington [76] reveals, that in a given class of upper society, the greater the man's success in life, the more children he tends to beget. Carr likewise shows how the frequency of patenting declines steadily with growing age, that boyhood and environmental influences are markedly important, and only one person out of 551 in the population takes out a patent (or more) in the course of a decade. Sorokin finds inventors preëminent for long life.[171,1]

Hornell Hart [159] and seven students have made an excellent inductive and statistical study of the biographies of 171 prominent English and American inventors of the last three centuries. These were found characterized especially frequently by persistency, intelligence, insight, energy, eccentricity, timidity and sensitiveness. They were inspired, it appears from their biographers, chiefly by the pleasures of invention and successful mental functioning, and by rendering service, desire for gain being rarely mentioned. We think it would have been oftener ascribed to those heroes, were it not a sin according to the official Christian ethics. They were noted as actually receiving wealth in 28% of cases, and fame or recognition in 37%, with better ratios for modern finders. They make repeated inventions, but are held back especially by poverty, ridicule, lack of equipment, initial failure, etc. Their education was usually very good, and a quarter had professional fathers.

---

[76] Ellsworth Huntington & Whitney: *Builders of America,* pp. 32, 248, 251, 254. Data from Harvard and Yale graduates and *Who's Who* personalities. And Huntington: A Neglected Tendency in Eugenics; in *Soc. Forces* 12:1-8.

Amy Tanner [256] concluded that genius or talent has no great relation to fame won, and that inventive power depends on manipulation, has been experimentally studied thru puzzles, very often manifests itself in youth, and is sometimes helped by an inferiority complex, as others have noted. E. With [257] pointed out that inventors (like artists) are never men born to wealth, and are rarely scientists and almost never scholars nor writers. Those are not their jobs. But among the inventors considered for the ship a large part have written books or pamphlets when their inventive business called for it, with no evidence of incapacity. Taussig's *Inventors and Money Makers* is a classic work, tho outmoded in its psychology, and like Hart's stresses the work's-sake motives of inventors [177] . . . Probably the newest good book in this field is H. S. Hatfield's *The Inventor and his World,*[161] by an English scientist and inventor, who entertainingly describes the personality and problems of the free-lance inventor in his country . . . The numerous and statistical writings of Dr Rossman, published in his *Psychology of the Inventor* [171] and in the *Journal of the Patent Office Society* (Bibliography 209.1-218), are the most valuable of all, containing accounts of their respective mental methods by a number of leading inventors, psychological analyses, and statistical data on their traits, motives, personal origins, occupations, obstacles, etc.

But as pointed out in our Principles 33 and 28, the inventors are not the only individuals responsible for invention. There are their very important backers. And the whole population takes some part (Prin. 28), infinitesimal with most, but of increasing value as their knowledge is more special and contact with the inventing closer. Probably workmen, foremen, salesmen, executives, consultants and patent attorneys have considerable shares in this way, that it would be a nuisance to try to acknowledge in the patents. Indirectly, thru building up wealth, ready purchase and maintenance of new articles that require intelligent users, and in endless other ways, the whole population takes part, and so do the geographic and racial factors that underlie civilization. (Prins. 2, 6, 16).

For further illustrations of all Principles relating to the inventors, we must refer to Chapter VII of our companion work, on the History of Ship Designers and Inventors. This displays especially the (indeed self-evident) key position of the inventor class, (Prin. 27), the trends of inventive method (Prin. 31), the non-necessity of science for invention (Prin. 4), the importance of the professional inventor (Prin. 17d), the Principles centralizing invention (15-18), the uselessness of the premature (36) and slow incubation (8). Our history, after telling how naval architecture became a scientific profession in 1860-80, leads us to

The Occupations of Marine Patentees

In addition to the aforesaid observations and surmises on the professional occupations of our notable marine inventors, the study of patent statistics gives additional evidence that the designers of ships are increasingly becoming the same as the authors of marine inventions, until today the inventing of the ship has passed pretty thoroly into the hands of trained naval architects and related professionals, to whom the ship is a specialty field of the universal engineering science.

Barnaby's [77] count of the occupations of the recipients of 292 patents relating to shipbuilding, taken out in England before 1852 and naming the profession or title of the inventor, revealed only 20 who were shipwrights or naval architects, compared with 80 who were styled Gentlemen, and a motley array of landsmen of all ranks. The pat-

TABLE I: OCCUPATIONS OF BRITISH PATENTEES FOR MARINE INVENTIONS

In Percentages of the Identified Totals (ftN *80*)

Occupation	On marine propulsion MacGregor to 1857	On marine propulsion 1912 and 1913	All marine classes, 1913
Presumptive insiders, total	**46.4**%	**63.** %	**68.** %
Naval architects		2.6	7.5
Ship or boat builders	1.7	1.3	5.4
Engineers	40.7	49.0	37.2
Ship owners	1.2	2.6	2.0
Officers of merchantmen, and pilots		3.3	6.1
Naval officers	2.1	.7	.7
Mariners	.7	.7	2.0
Inventors		.7	2.0
Other insiders		2.0	4.1
Presumptive outsiders, total	**53.6**	**37.1**	**33.0**
Mfrs. & land executives		4.6	5.4
Architects and builders (M—"architects")	.6	1.3	3.4
Draftsmen		0	1.3
Foremen		0	.7
Workmen (M—"machinists")	.7	8.0	4.0
Gentlemen	37.4	4.6	6.1
Peers	1.2	0	0
Physicians	1.7	2.0	2.0
Learnèd professions		6.0	2.7
Merchants (M—"tradesmen")	11.1	4.6	4.0
Women	.3	.7	.7
Misc. outsiders	.6	5.3	2.7

The following figures are absolute, the foregoing being percentages.

Total with specified occupations	670	151	148
Occupation not stated	160	3	3
Firms		5	5
Patents, instead of patentees	802	127	139

ents were so poor that Barnaby said he found no improvement worth recording, from the start in 1680 to 1810, except in the making of sheathing and ship's pumps. This last rather unmarine field included more than a third of the patents before 1800, which may be one explanation for the predominance of landlubbers in early marine patenting.

MacGregor,[78] compiling all the British patents from 1618 to 1857 on marine propulsion, found occupations stated as per table I, and many foolish patents. Seeking modern statistics, and finding few countries in which the occupation of the inventor is regularly supplied,[79] we have thot best to match the propulsion statistics with those of 1912, 13 and also consider for the latter year half of all the British patents for ships, boats and rafts and their parts, handling and propulsion, not including engines. The occupational groups in the earlier series show as many outsiders as insiders, but in the later only half as many. And we must presume that many of those assumed outsiders have really some first-hand acquaintance with the aspect of ships they were handling,—for instance some of the manufacturers, draftsmen, workmen and Gentlemen. Comparing MacGregor's figures with those of 1913, we note particularly the rise of engineers and naval architects, the decline to insignificance of patenting by naval officers, the decrease of those styled Gentlemen and of outsiders generally, the disappearance

[79] The occupation of the patentee is uniformly given today only in the Danish, Norwegian, Indian and N. Z. publications. But the latter have no shipbuilding industry, and the first 3 do not give the inventor when the patent was assigned.

[80] The number of patentees is somewhat greater than the number of patents, due to plural inventors (ca. 19% in 1912, 13). The percentages (except of the Grand Totals, as given in the last 3 rows) are calculated taking as a base number the total of patentees having their occupation specified. The 2d column is part of Class 114; the 3d is Classes 113-5, on the R hand pages only of *Abridgments of Specifications*. From both classes patents relating essentially to aircraft were excluded. Occupations were obtained from the *Specifications*, and cover all nationalities in the British office.

Table II: Occupations of Patentees

On inventions in all lines, in the Norwegian and Indian offices, 1923-4 [81,82]

	Individual Patentees	Engineers, Chemists, Electricians	Foremen and Mechanics	Production Executives	Learned Professions	Untechnical Occupations	Unstated Occupations	Firms No.	Firms, Per-cent. of row
Percent. of 146 assignable individuals (ftN81)	106	46	7	22	15	10	6		
Total number	154	67	10	33	22	14	8	53	26
Norwegian patents, number	69	27	9	18	4	6	5	31	31
To Norwegians	19	6	5	4	0	4	0	5	21
To Foreigners	50	21	4	14	4	2	5	26	34
Indian patents, number	85	40	1	15	18	8	3	22	21
To Hindus	12	5	1	0	1	5	0	0	0
To Europeans in India	9	3	0	4	2	0	0	3	25
To Patentees abroad	64	32	0	11	15	3	3	19	23
U. S. Inventors, per Rossman	710	60%	5%	24%					

[81] The first percentage (106) is over 100 because it includes unassignable individuals. From *Norsk Tidende for det indus. Retsvern,* whole issue of Oct. 8, 1923; and the Indian from *Specifications of Patents,* v. 102 from Feb. 18, 1924 on.

[82] Rossman's data are from a questionnaire of unstated date and percentage return, to American inventors, presumably properly weighted in favor of the more prolific patentees. Numerous questions such as that of insiders vs. outsiders, and the decline of the "Yankee inventor" are well and statisticly discust in Chaps. 3, 9 and 10, respectively on the Classes, Occupation and Motives of Inventors. *The Psychology of the Inventor,* our note 171. Cf. the same author's Do Engineers Invent, and Why do they Abhor the Label "Inventor" (N 210), and Engineers as Inventors (N 211).

of the nobility, and the persistence of a sizable contribution from architects, probably from their engineering affiliations.

For comparison with these data we have compiled others in Table II on all kinds of inventions, patented in 1923 and '24 in the Norwegian and Indian offices.[81] The top row of figures, being in percentages of the assignable individuals, is comparable with Table I, and shows the same overwhelming preeminence, first of the technical men, and second of the production executives, with outsiders taking perhaps around 15% of the patents. They would be represented chiefly in the Learned professions and the Untechnical occupations, altho often such men are competent in the field in which they are inventing. On a different basis we have excellent data for modern U. S. from Rossman's study of inventors.[82] Along with still greater preponderance of the technologists, he finds 2% to be professional inventors, 1.7% patent lawyers, .7% merchants, as many farmers, and 5.9% miscellaneous.

However in all studies of invention measured thru patents, one must remember as in Chap. II that a patent is by no means equivalent to a unit of invention. A patent may represent a grand invention, or a little one, or nothing but old stuff, folly or chicane. That few patents are ever used or useful is reflected in MacGregor's finding that one-third of the marine patents of 1852-7 were allowed to become void after six months. Thumbing many patents has given me the impression that those of outsiders are much oftener impractical than those of insiders. They may be too revolutionary. (Prin. 32). Supposing that what we are now desirous to measure is weight or value of inventive contribution, our own statistics are probably further inflated in favor of the outsiders, by the fact that some of the patents have been assigned to a corporation before their granting, and that the inventions thus removed from our count presumably average better than the rest. Many were made to order by engineers and chemists in corporation laboratories, and if not perhaps useful, at least represent the most intelligent effort, at either creation or forestalling. The extreme right column shows that in Norway and India such patents are 26% of the enlarged, grand total.

One means for evaluating patents is by the good supposition that the farther the inventor goes afield in seeking patent offices, or the more recondite a foreign office be from the standpoint of his own country, the more valuable should be his patent.[194] Accordingly we find the two Outsider classes getting about twice as large a share among the natives as among the foreigners, alike in the Norwegian and Indian offices. Doubtless the disproportion would be greater, could we know the paternity of the inventions patented by firms. Among the patentees in shipbuilding in the Swedish office only 64% of the applications of natives were granted, vs. 90% for the foreigners.[83]

Still another means of seeing who make the most valuable inventions was used by Hart [159,196] who looked up the 171 inventors mentioned in Kaempffert's *Popular History of American Invention,*[285] and found that about 3/5 were old hands in the work affected by their inventions, about 1/5 were professional inventors, and another fifth were amateurs or outsiders to the industry.

It may be noted that few practical inventors, in the marine as in most fields, have been pure scientists. There is no reason to infer lack of ability among scientists—ordinary invention is simply not their job, by definition of their occupation. In marine history as elsewhere they have invented when their science was wanted in practice.

The professional inventor, at least the free-lance, miscellaneous one, unconnected either with a corporation or with one industry, has been conspicuously missing in the history of the ship, as from valuable progress generally. A number of our protagonists might be called professional inventors—Evans, Fulton, Ericsson and Flettner, and devoted themselves to the ship after making their first big success there (save that the great success and fixation of Evans were in engines and flour mills. The man with ingenuity and liveliness, but no profound knowledge nor persistence thru years, is evidently unequal to or unattracted by the watery element, whose life and problems are so different from those of the land. But in other lines too this classic type of inventor gets more notoriety than success, while in fields that demand scientific training, such as chemistry and electrical engineering, he misses fire or is missing as uniformly as with the ship.

Who Make the Revolutionary Inventions? (Prin. 32)

So much for the common run of inventions. As to those of revolutionary importance, the principle has been propounded and well supported by Kaempffert and others [84] that these cardinal inventions are due to men outside the occupation affected,[85] and the minor, perfective inventions to insiders. It has been certainly true as to the ship, that the striking innovations of known (and therefore modern) origin have all been made by men not professional sea-farers nor owners, nor even builders nor designers of ships, tho they often were led by their

---

[84] Seaton cites F. P. Smith, Bramah, Woodcroft and other marine inventors, and says that even in our own day we receive many valuable marine inventions from men outside the engineering profession. (Instn. of Nav. Arch., *Trans.* 60-54, 1918. W. Kaempffert vigorously presents the outsider thesis. (N165, 201).

[85] Those who claim this always make use of the fact that most past inventors can be assigned to any of several occupations. Thus they call Leonardo and Fulton artists, altho each of these considered himself an engineer, and was.

inventions to become such. Yet they were, if successful, almost always men with extensive connexions with seafaring, like Bouguer and the other founders of Naval Architecture, I. K. Brunel who had been building railroads till he designed the *Great Western,* but whose famous father had been a naval officer, Ericsson who had just put thru a marine engine, and Flettner with his youth on the Rhine and the Roaring Forties. If Fulton and John Stevens were quite unmarine at the start of their steamboat work, they continued experimenting long enough to qualify. In no case do we find a good invention for the merchant marine being made by a sheer landlubber, nor by a sailor, and scarcely by a ship's officer. To reform the ship basicly it has always been necessary to depend on some outsider, who yet had sealegs on him, knew his footing on water, while his head could see into other spheres of life—Mathematics, Physics, shore Engineering, Aviation—to draw thence precious suggestions. The professional devotees of the ancient, well-loved and piously reverenced ship are forever perfecting her, but had as lief capsize her as turn her upside down.[86]

However, mere revolutionary *ideas* like those of a steamboat, a propeller, an iron ship, or turbine or rotor, are of little value to civilization, as we have seen, because they occur to several men, so that any one is dispensable, and because their bringing to fruition will require a vast deal of work, to bring both the idea and the people to practice. This gradual and painful introduc-

[86] Usher well points out the value of day-dreaming, wishful thinking and emotions, which enable the inventor to escape from the shackles of logic, with its false traditions and authorities and unsuitable habits. "The disregard of 'logic' and 'reality' promptly removes all the inhibitions of the over-disciplined voluntary mind. The utilization of the entire field of memory furnishes a richer array of materials than is afforded at the level of full consciousness," which would have rejected great masses of data as useless, that might have served. (N 179, pp. 30, 31). The same liberation applies to the outsider. Cf. FtN *36.*

tion, in which the semi-outsider predominates, is one really important part of the service of inventors. Their other important work is the continual betterment of established devices, almost always by insiders.

We have seen the ship modified far more by evolution than by revolution—by minor or gradual, even imperceptible changes rather than by novel large principles. So the insiders who did almost all this perfecting have contributed far more to the building of the noble fabric of the ship than have partial outsiders, altho the latter win the greatest honors.

With regard to inventions in general, the reason for the rule of revolutionary inventions coming from outsiders is first of all that when an industry is by hypothesis just being started, it can hardly have any insiders. Also when the industry is still small, or even large, outsiders still outnumber the insiders many-fold, and simply on this numerical account might contribute a greater number of inventions, than insiders who were per capita many times more prolific in ideas. Next, Sweeney's [175] data on the occupations of numerous fathers of eight grand inventions suggests to us that there is a tendency, generally as well as on the waters, for the successful inventors to arise from the occupations possessing the needed knowledge. Physicists, electricians, chemists, engineers, and mechanics have been constantly conspicuous in the quests of the inventions especially needing them, as the thermometer, electric light, photograph and airplane. Numerous farmers and agricultural implement manufacturers helped invent the reaper. The typewriter and sewing machine, which demanded only mechanical knowledge such as is common, had a highly heterogeneous crew of fathers . . . In hardly any case do we find invention proceeding from the industry unrelated in methods to and subject to replacement by the novelty. Coal-tar dyes were not invented by indigo-planters, electric lights by gas engineers, nor autos by liverymen; for reasons that are obvious and compelling.[87] In fine, possession of needed knowledge, concentration of attention, and motive for action, seem to have been three great determinants of who shall invent a thing; and these give the best per capita chance to insiders, or the nearest approach to insiders in the case of a new industry. But the principles of overwhelming numbers, and of freedom from custom, give most of the revolutionary inventions to outsiders.

---

[87] One of our greatest inventors, Fessenden, went much further, with the rash statement: "No organization engaged in any specific field of work ever invents any important development in that field; or adopts any . . . until forced to by outside competition." Many instances and documents are cited. (N90, pp. 103, 4).

CHAPTER V

THE HARD STARTING OF FUNDAMENTAL INVENTIONS

Marine Inventors and the Patent System

It is not surprizing that we have found most of the known inventors of the ship to be ship designers or builders (and for earlier times we have presumed the same); because before the patent system of the last couple of centuries there was no way whereby anyone else could hope to carry out his idea and be rewarded for it. (Prin. 27). Nonetheless the contribution from rank outsiders, not including engineers and mariners, we have seen has been diminishing almost to the vanishing point. Any fool may take out a patent; but there are so many steps to pass between even a good marine invention being patented, and a net reward being cashed by the inventor, after the assistants, experimenters, lawyers, builders, capitalists, and the competitors who appropriate the idea or fight for it, illegally or lawfully, have all drunk their fill from his invention, that the lay inventor will best protect his happiness by deepest pessimism.

Even insiders have made no notable wealth from patents on ships or their parts, in any cases that we have been able to learn of, save presumably from Sickles' cut-off (?), Parsons' marine turbines, Isherwood's framing patent, the Sperry gyroscopic compass, Flettner and Oertz rudders, outboard motor, various marine boiler patents and perhaps some other recent inventions. Most of these are out of our selected course of ship descent. What are always called the "patents" of Fitch and Fulton were valuable encouragements to invention (tho they made little money for Fulton or his heirs and none for Fitch); but they could not have been granted under modern law, and were not patents in the modern sense, but *concessions* for the navigation of certain best waters by steam. Our federal patent system almost started with the noteworthy but inauspicious grant of four patents on a summer's day of 1791, to Fitch, Rumsey and Stevens on steamboats conflictingly defined, and to Read for a tubular boiler. . . . Ericsson received no net return from his propeller patents, but a £4000 gift from Parliament. He refused to protect his *Monitor,* and while he took many other patents, he wrote of one of his inventions "Like all other original ideas, entitling me to fortune and honor, the one in question will also be stolen from me".[88] I. K. Brunel exprest bitter disgust with the patent system,[89] in that period of wide protest against it.

In the marine field public rewards to inventors are less needed than in some others, since the discoverer has no alternative of secret use. The size and movement of the ship and the irresponsibility of the crew make secrecy impossible, except as to her lines, and statistical measures of efficiency. These are unprotectable by patent, and often kept secret. For the rest, the impracticability of secrecy would encourage patenting; but authorities say that there has been less patenting in ships than in other fields . . . . We shall speak at the end of this chapter of the declining disposition to take patents in all fields, and the needed study of the matter.

Financial rewards for invention may be good social policy even if non-existent, so long as worth-while but deluded inventors continue to expect them. Still the patent system hardly seems justified by the history of the ship, unless in the immediate past. Its benefits are ob-

[89] As consuming the energies and blocking the progress of inventors. Quoted in Church (N 88) 2: 240, 1, as of 1856.

vious; its hurts, in its vast consumption of wealth and inventors' efforts, its obstruction to combining good new ideas, its subjection of inventions to the whims and fortunes of individuals and companies, its taxation of novelty (by royalties), need to be called to mind. The writer cannot pass judgment upon the general problem of the patent system; but would point out two of its likely desirable alternatives. One is a public grant of money after the worth of the invention has become clear, as in the case of the propeller, from Parliament, and Sepping's docking invention, from the Royal Society of Arts. Another is the support of the inventor, and wise direction of his efforts, by the state or philanthropy, not *ex post facto,* but when he needs help and guidance most, viz. while making the invention. We have recorded such help in the cases of Hoste, Lord Stanhope, Beaufoy, Froude, Taylor, Flettner, the perfecting of the propeller by the British and French navies—and indeed the navies and other public services and the scientific institutions now and in the past have been constantly and successfully employing this principle to invent as well as to discover. We will state this despite Fessenden's supported claim [90] that government boards have been flat failures at revolutionary, original invention, whether directly or thru agents, and little appreciative of the offerings of independent inventors.

The Long, Long, Useless Stage (Prin. 8)

The need of such non-commercial support appears most strongly for the class of inventions which share with scientific discoveries (almost always so supported) the traits that their practical benefits, tho likely great are probably remote, 10 to 50 years in the future, and quite uncertain as to amount, kind and beneficiaries, and probably unrewardable in practice by the patent system, whatever the patent laws may pretend. This class of inventions is the *fundamental* ones, the basic starts in wholly new directions. (Prin. 8) Let us list in Table III those in the history of the ship for which sufficient data are at hand, with the dates of the first work done on the invention, first commercial success afloat, and important use, and then set down the two intervals, and the averages.[91]

TABLE III: MATURATION OF FUNDAMENTAL MARINE INVENTIONS

A Inventions ashore or afloat	B First plan	C Comr'l success afloat	D Impor'nt ant use afloat, ca.	E Intervals, approx. C-B	F D-C
Steamboat	1661	1808	1815	147	7
Iron ship	1777	1829	1840	52	11
Propeller	1746	1836	1840	90	4
Surface condensation	1765	1858?	1870	93	12
Oil fuel	1860?	1861?	1905	1	44
Turbine, steam	1st Cen.	1897	1905		8
Diesel engine	1882	1903	1908	21	5
Motor vessel	1000	1906?	1910	100	4
Contrapropeller, fixed	1864	1910?	1920	46	10
Turbine, geared	1629	1911	1920	282	9
Stoker, mechanical	1816?	1921?	Not yet	105	
Rotorship	1923	1926	Not yet	3	
		Geometric means		41	9
Average Intervals:		Medians		90	9
Cf text and ftN93)		Arithmetic means		86	11

Some statisticians will say that no average is permissible, of such disparate and arbitrary units (or disunities) as inventions. But I think such averages are allowable, to give some conspectus, some simpler grasp of the subject, provided no accuracy be claimed for them. Altho the number of cases is small too, the average Interval seems

[91] As to averages cf. ftN*93* and pp. *23* & *29-32*. The Alexandrian date was not counted, to avoid an abnormally large item in support of our case.

to show well our 8th Principle: how slow progress is before the day of first commercial success, compared to after it, as we have discust in the case of the steamboat. (Cf. also Prin. 14.).

The reason is of course that there are few people willing or able to work on an invention whose past has been one of uniform failure, and whose future offers no certainty but that of expense, probably great and lasting long, unless discouragement supervene earlier, in which case usually every cent put into the invention is lost.[92] I have looked into the histories of 19 most useful inventions introduced in the 25 years before 1913, selected by vote of *Scientific American* readers. The average[93] intervals were: between when the invention was first thot of and the first working machine or patent, 176 years; thence to the first practical use, 24 years; to commercial success 14 years, to important use 12 years; or say 50 years from the first serious work on the invention, to important use from it. Again, in the study of the most important inventions of the last generation before 1930, for Prof. Ogburn and *Recent Social Trends*,[94] I found a median lapse of 33 years between the "conception date" corre-

---

[92] H. S. Hatfield says: "The most frequent cause of the downfall of the brightest of [the free lance inventors] is . . . . the pioneer invention. He conceives a fundamentally new idea, which he feels is assured of startling and immediate success." He gets in deeper and deeper, and "It is a notorious fact that many if not most of the pioneer inventors to whom the edifice of modern technology owes some of its chief pillars have died in poverty, or if alive, are receiving absolutely no reward at all for the incalculable benefits conferred upon industry by their labors." N197, pp. 111, 2.

[93] The geometric mean was chosen as more suitable than the ordinary average, the arithmetic mean, which was as much as 3-fold larger from giving more weight to the larger cases. Fuller information would increase the intervals. Two Greek dates were changed to 900 years later, so as to close up the pause of the Dark Ages. The median intervals were 210, 33, 12 and 14, total 269, and the means 422, 70, 47 and 14, total 552. The question of the *Sci. Am.* was: What are the ten patentable inventions introduced commercially in the last 25 years, which are of the greatest practical success and usefulness? *Sci. Am.* 109: 163, pp. 243 & 352, 1913. The 19 inventions receiving most votes were used above; 20 were intended, but the 11th, "calculating machine" was dropt because of difficulty of definition.

sponding to the 2nd above, and the date of commercial success. A third of a century or a half—it's a long time for a person or firm to work. And it is only an average, with no sort of guaranty; and it is an average for the most successful inventions only. No wonder the gestation period of inventions is long, and could and should be shortened.

This period provides one ready means of profesying inventions (Prin. 14). One can simply predict the distant future use of inventions already started but hardly effectuated.[95]

We have claimed elsewhere that great ideas, the starts of grand inventions, are things of very little value, being both premature and redundant, and that what *is* of value is great quantities of hard, developmental work, to bring the grand ideas to fruition.[275] We are claiming now that they could be ripened considerably quicker, if such effort and funds as seem socially justified were spent on each case. To be sure, the ripening of each invention depends on the maturity of others (Prin. 6a)—the propeller and the turbine had to wait upon better steels, metal-working and other factors cited. But could not the metallurgy have been improved faster, with benefit to a host of inventions? A general policy of apportioning inventive effort according to social need would produce a widespread speeding up, in all lagging sectors of the advancing battle front of technology.

The incubation period of inventions is long because the commercial motive is crippled in the earlier listed period, potent in the later, and we act as if we trusted it to actuate both periods. Suppose we left the medical and all

[95] To be sure, the infantile death rate of inventions is very high, so that it is no easy matter to say which of the latest projects will succeed, as the prevalent blunders of the admiral and historian Preble, regarding the latest ship proposals of his day, ruefully remind us. Cf. Ogburn N 237, pp. 164, 5, and Chap. VI of *Inventing the Ship*. Prediction is much safer and still useful when broadened according to our Prin. 34, to allow for alternative means of reaching the same goal.

other scientific discoveries to be made only *if* the discoverer would be commercially repaid, after ten years of patent struggles, or if he deludedly hoped he would be, or if he were a philanthropist or a lover of science, with the wealth, time, ability and will to devote his all to scientific discovery? i. e., suppose all our present support for scientific research thru universities, foundations, special funds and government bureaus reverted to the meager backing for science two centuries ago, plus commercialism? Will anyone think that science would prosper? Yet that is just what we do in the case of fundamental inventions, which differ (for our purposes) from science only in that their practical utilities are more foreseeable, and their development more expensive, since they require construction on a large, not a small, laboratory scale.

Take for example the jet propulsion proposals which have been made from time to time for 252 years—we have elsewhere described the projects of Boerner and Paxton, and quite a few engineers are expecting great discoveries along this line, for ships, aircraft and rockets. The American and British navies, and a French bureau of which anon, made some experiments, but otherwise the whole costly field is left to the bankrupting kindness of occasional individual inventors, so that while the centuries drift by the idea remains unplumbed, with advances made from time to time, but never enough to overtake the constant progress in the commercially safe, orthodox, abundant types of propulsion and hull form. (Prin. 15)[95.1]

[95.1] A. E. Seaton, a marine authority, recalls as great losses, from lack of modern research, the cases of the direct-acting engine, screw propeller, surface condenser, compound engine and turbine. In other cases he cites, such as Yarrow's balancing of reciprocating engines, research was used and success won in a reasonable number of years. The temperament and training of the ordinary engineer are not suitable, he says, for research, and he is too busy. Moreover, we would add, he has not the often very expensive equipment needed, such as model-towing tanks, and governmental authority to collect data, and demand information on the results of private tests, or

Or take the ladder hydroplane, or any other of the latest marine inventions; or for a striking case, that of the voice-operated writing machine.[96] The great Fessenden invented one in 1907, and Jn. B. Flowers in 1916 made one that would operate, and a typewriter that would write the vowels when spoken to; but since then I have heard nothing of the contraptions. How many more decades shall we have to wait for this glorious invention, the mechanical scribe, just because Everybody's business is nobody's business? L. C. Smith was backing Flowers; but there is no very good reason why a typewriter manufacturer should pour out money on an off chance of some day creating a successful machine highly different from a typewriter, that would largely supersede typewriters, and probably yield all its profits to later comers, by the time a nation had been brot to use a new alphabet and spelling. So it was with five of the greatest latest inventions, the thermionic valve (vacuum tube) for 30 or 180 years,[97] photoëlectric cell for 40, picture

---

the right to make tests on private ships or equipment. Seaton proposes a list of 34 marine fields needing investigation, such as might be carried out under his British Marine Engg. Design and Construction Com'ee, using part of Parliament's million-pound wartime grant for scientific and industrial research in all lines. He also proposes to search thru the junk-yard of forgotten inventions, from which the corrugated furnace was rescued. "It was because so many of the old patents were mere ideas only, and not really inventions, that they were never proceeded with . . . . With due research . . . the knowledge and methods of today can make some of 'their dry bones live' in a way that was not possible formerly. Nor could they be useful for purposes which did not then exist, whereas today they may be a great advance to satisfy modern requirements." (After some of the many changes constantly occurring under Prin. 2.) Some notes on the Importance of Research in Marine Engg., (ftN *84*), 1918.

[96] Fessenden N90, p. 134 and U. S. Pat. App. No. 358,078, Feb. 18, 1907. Flowers' inv. in *Sci. Am.,* Feb. 12, 1916, p. 174. Predicted as the phonoptograph in 1892 by Max Plessner: *ein Blick auf die grossen Erfindungen des 20. Jahrhunderts;* 93 pp., Dümmler pub., Berlin. Recent fundamental marine inventions will be found in Chap. VI of our *Inventing the Ship.*

[97] Du Fay in 1733 noticed the quicker loss of an electric charge by a hot body, and Edison patented in 1884 his vacuum tube hot cathode rectifier (U. S. Pat. 307,031).

telegraphy 80, television 40 and stereoscopy for 100 or still, until they were taken up by the great Bell and General Electric companies. These corporations, with their almost monopoly positions and international ramifications, more resemble socialist governments than competitive firms, and have in recent years been acting appropriately, for which we may thank them. They have been persistently and effectively developing both pure science, and fundamental inventions, like television, the photoelectric cell, talkies, stereoscopic movies, and clear fused quartz. But there are few other companies that occupy such a monopoly position, and respond with the wide and long views it makes appropriate. One might mention for honors the Westinghouse, and Eastman Kodak companies, and recall as we did in connexion with Prin. 17d, that a number of trade associations are doing research, with the proper all-inclusive backing, but with no great funds nor often a program in new fundamental inventions and science.

Endowment Needed for Fundamental Inventing

There are many government divisions, like our Bureau of Standards, which spend millions to accomplish big results in testing and improving invented processes; but they do not try to make any inventions of a fundamental character, unless for purposes of war. Other governmental bureaus, and universities and foundations, do so much for scientific discovery that its main support is thus organized; but again little for invention. Neither government nor philanthropy will sully its pure hands with inventions from which someone *hopes* to make money. Neither will money-grubbers, when the hope of fingering some net returns within 15 years is as derisory as we have seen it in the case of new fundamental invention.

Yet examples can be found of about what seems called for.[98] In this country the eminent inventor, F. G. Cottrell's foundation the Research Corporation, and the Engineering Foundation, help varied researches, but not much invention, save that the former continues the development of the electric precipitation process. The Guggenheim Foundation and the American and Cleveland Rocket societies have financed aviation invention, and that very promising and difficult fundamental novelty, the high-power, exploratory rocket. In France is the Office national des recherches et inventions,[99] which along with familiar testing, research and military functions, may help any French inventor to work up his device (or counsel him to drop it), resolving questions of detail, bringing in competent people, seeking out the prior art, getting constructors and industrialists to try out the novelty, and helping protect the property rights of the inventor. In return it takes a share in the patent, and can protect the same at law, being a corporation. While most of the inventions told of seem not in novel fields, they have taken up 3-dimensional movies, photosculpture, the windmill boat of Constantin, and a sort of jet propulsion. The organization began in 1915, for war purposes, and received during the war 44,976 propositions, of which they took up 1958 and worked out and made ready for use 781, a much better record than the American board;[90] but they seem to have received plans from army officers. The state budget for 1932 was only $110,000, plus something from gifts, collaboration and patents.

Such organizations, to shorten the gestation period of the great inventions of the future, by working on ideas that promise great commercial utility, but not very soon nor surely, or which for any reason are not likely to be rewarded thru our patent system, are recommended by Ogburn [100], Vaughan and Watson,[101] and by Fessenden,[90] despite his scathing strictures upon boards in the past.

For his own experience with the patent system had been most unsatisfying—a life-time of brilliant achievements in electrical invention, with patents taken or applied for, and fought over, yet not a dollar of return from patents, he says (until the very end of his life).

Still other means may be found to approach the same goal of bringing new fundamental inventions from the clouds down to functioning reality. We mentioned some pages back lending the service of ship model tanks and wind tunnels to private inventors. But at least the wind tunnels charge so much, $700 for testing one model, that they largely thwart this purpose, while standing idle most of the time. Such a situation of large unused overhead calls for variable charges based on "what the traffic will bear," or else not attempting to make the project self-supporting, as with airways and libraries.

Ravenshear [170] proposes to grant longer-lasting patents for fundamental than for perfective or detail inventions. Perhaps; but the measure could bring nothing but a distant, doubtful, ex-post-facto reward, and would much increase our burdens of legalism. Some countries adopt this principle by limiting the life of a subsidiary to that of the basic patent, and our own follows Ravenshear's principle in rare Congressional extensions of an insufficiently rewarded patent. At various times we have seen governments and societies awarding prizes for inventions. But if they are ex-post-facto and rare, like Parliament's £10,000 to Cartwright for his textile inventions, and £20,000 to the inventors of the propeller, they do little to stimulate and guide invention. Prizes offered in advance come nearer to meeting the need, like some that helped the early automobile and aircraft builders, and brought out Stephenson's locomotive, with Ericsson's good runner-up. Nothing was being done on aviation in England till the *Daily Mail* began offering prizes; and when the Air Ministry some years ago offered

£50,000 for a viable helicopter, there were 15 or 20 competitors. But still we lack a useful helicopter, that indispensable flying machine of the future; and how shall we get it soon unless someone pays in advance to make this most expensive invention? Millions have already been spent on helicopters, with some subsidies such as we propose, by our own and other governments, but too half-heartedly.[101.5] If the inventor does not have means from *some* source *before* he starts to work, he cannot contrive even an improved toothpick.

Babies Needing Foster Parents

It is much that we are missing thru thus leaving fundamental invention for the most part as unpaid-for as science was in the Middle Ages. There is for instance the telharmonium, the supremely beautiful musical instrument of the future. Or we might get some equivalent, embodying the same principles of making tones electricly and syntheticly, hence with completest control, variety and range, instead of depending on the quasi-natural, mechanical-vibration notes from voices, violins, etc., which long ago reached a stage of inventional stand-still (Prins. 9, 12, 13), so that there seems scant hope of remedying their many faults. A telharmonium would be immovable and very costly, adaptable only to broadcasting, and to exploit its new powers, especially of playing in the just or "natural" intonation, instead of the conventional false, tempered intonation, will probably call forth the recomposing of all music. Since Cahill got it going commercially in the first decade of the century,[102] but had to abandon it after losing hundreds of thousands, no one seems to work on it unless perhaps the Bell people. Naturally—everybody's business is nobody's business. Yet in the meantime the advance of other sides of life has made success much easier—there is more wealth and musical appreciation, there is radio broadcasting, and there are 4 alternative means (Prin. 34) for creating the tones (the heart of the invention), which Cahill made by hundreds of expensive dynamos:—movement of tone films, magnetized disks or wires, phonographic records, and vacuum tubes, the first 3 being imprest with waves either by natural sound or artificial process.

There is the problem of a scientific alphabet and orthography, and instruments for writing it, coordinated with the mechanical stenographer. Since *reading and writing* are coming to be the chief occupations of the more important part of our population, the importance of these problems is overwhelming, far beyond that of ships, for instance. There is the optophone or phonopticon which reads ordinary print out loud,[103] and could write from it too, or set type. It has been operating after a fashion for at least 20 years; but who will perfect this boon to blind and sighted? Possibly someone who is blind to the odds. There are problems of getting more electricity direct from heat, or from coal in the mine, from sun power, from the temperatures of tropical sea-water—Claude spent an uncommercial million just to get this idea successfully started.[104] There is under-ice navigation, seeing thru fog by infrared, all the novel ships, and even old problems that inventors have persistently attacked with small means and never conquered—the ship brake, unsinkable ship or cabin, non-capsizable life-boat. There are amateur photography in colors and 3 dimensions, the active picture, television, internal combustion turbine, cotton picker, food syntheses, artificial fabric clothes without weaving or sewing—any number of basic new starts. All these are civilization's babies—born, but useless as yet. Suppose we abandoned all babies as not being "good commercial propositions?"

To be sure, our Principles 14, 24-6, 34 and 36 seem to claim the inevitability of invention at its proper time and not before. But all Principles are only tendencies, variable in operation within certain limits; and it were highly desirable for the great inventions to be set to work near the beginning of their possible birth period and not later, as we found in the cases of the steamboat, propeller, surface condenser, tube boiler, rotorship, and all the sailship inventions. Moreover, if we worked all along the line to develop fundamental inventions, it seems probable that the general inventive progress of civilization would be speeded up, fostering all contrivings both basic and ancillary.

THE INDISPENSABLE BACKER (Prin. 33)

Inventors never are men of wealth, save the few who got it from their own creating;[257] it is quite distinctive trait, and reminds us of their fellow creators, the artists. Hence the importance of the backer, whether an individual, corporation or government, who must always be called in sooner or later. Formerly it was later, commonly to exploit the invention; but latterly he appears sooner, to aid in development, pay for the first working out, and call in the first place for what should be invented. Hatfield [197] finds many parallels between the inventor-patron relation and the wife-husband one—the inventor creative, absorbed, emotional, timid, suspicious, helpless, often with an inferiority complex, and the patron clever in law and finance, having a nose for the problem that can be profitably solved, and out to keep the inventor in his place. The relation between the two needs institutional and state control, with limitation of permissible contracts, just as with marriage and divorce, this inventor Dr. Hatfield suggests, for one reason to protect the life of the offspring. The patrons are often accused, by inventors like Hatfield, Jenkins and Fessenden, of stifling the creative power of the men they hire, by fixed pay contracts and by routine. But they give very much in exchange. It seems likely that the highly varying inventive rates of the different regions [194] are due largely to the differences in the enterprise of their enterprisers.

The Obstacle of the Users (Prin. 29)

We shall need little argument to establish that the acceptance and introduction, and therefore the contriving of inventions, is gravely checked, and doubtless the direction of inventing somewhat deflected toward familiar channels, by the conservatism of the users of inventions, as well as thru the inertia of their makers and their patrons. Hart [104,1] gives many and striking illustrations of positive opposition to inventions. We recall many cases in marine history where an excellent invention was received as slowly and grudgingly as if it were a socio-political reform. The jib and other fore-and-aft sails, the rudder, steamboat, screw, high pressure, surface condensation, compound and triple expansion, improved rudders and rotorship have had to fight their way. Only the compensated compass do we recall as meeting what might be called an immediate acceptance. The causes of the consumers' resistance are various, and well discussed by Hart [104.1] and Stern [104.2]—unacquaintance, stupidity, all the factors in conservatism and selfish vested interests, like those which fought the steamboat, iron ship and Flettner rudder, also patent royalties partly blocking the use of inventions, tho they probably stimulated their making, and the entirely proper obstacles in standardization, invested capital, and the workmen's capital of acquired, specialized skill, which all we discust under Prins. 19 and 20.

Often opposition comes from notable men in related lines. But we should not sneer at authority on that account. For as Hart says, most projects are bad, and as to the newest fields there *are* no authorities . . . . Often an elderly authority's reputation has outlived his flexibility of mind. Then too, the mistakes of the pontifical, as of Dr. Lardner condemning the transatlantic steamship about 1838, or Simon Newcomb airplanes, are remembered with special frequency just because they are striking exceptions to the scientists' usual correctness, and becoming parts of the mythology of the great inventions, are misquoted. Sympathetically interpreted from their original words, those scientists were not far wrong.

As in the case of the original and the current-actuated rudders, so in general, we think we see a secular shortening of the period of introduction, which would reflect largely or chiefly a loosening of most of those various consumers' braking forces. This is confirmed from col. F, Table III; and we do not know of one modern case of an important marine invention being delayed from general adoption more than ten years after it reached the stage of practicality for general use. Assuredly not all shipbuilders and owners are progressive, but there are enough such to give a good invention a start, and the rest

---

104.2 Bernard J. Stern says that the vested interests are not only economic, with large funds for propaganda, but also "that less tangible vested interest bound up with the struggle to retain status, reputation and prestige . . . . The vested interest of personal self-fulfilment functions thru the attachment of an 'emotional tone' to the existing process, which makes devotion to it and resistance to any substitute for it in the form of an innovation, unreasoning and unreceptive to any scientific proof of the latter's superior effectiveness. The older form is suffused with a value and meaning which it does not actually possess; loyalty is justified." *Social Factors in Medical Progress,* N 174, pp. 11, 12.

will soon follow suit, as their degree of dessication and other circumstances will allow, one after another yielding to the increasing pressures of proof and new custom. As a corporate prince or Maecenas of inventive progress we must list the Hamburg-American Line, which accepted for full-scale trial Flettner's rudder, airplane-ship and rotorship. Other shipping firms of the German Empire and Republic were very progressive, and made quite a contrast with most American companies, not to mention with the 1700s or Viking days, or Egypt which hardly changed the hull form in 2000 years.

Principal causes of the growing buyer's progressivism must be not only the modern general enlightenment, but especially the intensive education, and selection before training, of the *few* who now decide large matters of shipbuilding. The growing size of ships and shipping companies has concentrated responsibility upon an able few who know their specialized business, more than previously, and are largely engineers. Where we do not find this centralization or Spencerian integration of choice into the hands of a select few, no great improvement is noticed—cf. the stupid third class passengers who would not embark in a motorship without a funnel, and the silly first class passengers who demanded at least two funnels.

CHAPTER VI

THE DECLINE OF PATENTING, AND RECOMMENDATIONS

The Tendency to Take Patents is Declining

The impression is reimprest by one writer after another that inventors are swarming more and more to the Patent Office. Yet the opposite can be proved with ease. Since about 1885 American patents have been increasing only in proportion to the population (cf. Prin. 27), no faster, despite all the incitements from increasing industrialism and civilization. We were granting a patent yearly for each 2675 inhabitants in 1883-7, and 1 per 2691 in 1925-9, since when the Depression has reduced applications by a third. (See Graph 5.) The few patents granted to foreigners do not matter much. In England[164] patents have been decreasing rapidly since 1910. German patents per capita increased at 24% per decade in 1877-1913, 17.5% per decade in 1913-26, and 17.2% per decade along the straight-line trend of 1920-33. In all countries, Dr. Sanders tells me, except a few like Greece which have not yet attained industrialism, the same slowing up of the patenting rate is to be seen. This internationality of the phenomenon rules out one possibility, that the check to patents might be due to arbitrary changes in patent legislation.

The American patenting rate should have vastly increased since 1885, by all the logic of Principles 2 and 11 (the accumulation of inventions and other novelties on which to base new inventions), 15 (the growth of wealth and industrialism and therefore of the production of invented goods), 16 (fountain lands), 17 (all the specializations except the international), 18 (large and leaderly firms), 21 (rapid growth), 22 (increasing population), 27 (the increasing proportion of engineers and such people competent to invent), 28 and 29 (the general educational level), 31 (the improvement of inventive method), 33 (better backers for inventions), and 4 others, making almost half our list of Principles, with few on the opposing side. Particularly weighty should have been the growths of industrialism and of technical higher education. We have cited the vast growth of collegiate engineering students; their ratio of increase in 1890-1930 was 39-fold, or 20-fold that of the population. Surely that should have brot an increase of patenting. The influence of industrialism, wealth and general education, toward the encouragement of patenting, we have tried to measure by correlations of the 48 states;[105] and found rather strong relations. And the range of patenting was 9-fold (per white adult male) between Arkansas and Connecticut, or 4-fold between Connecticut and the Dakotas. If any such patent responsiveness to industrialism and other measures of Civilization has prevailed in the last half-century of time, as prevailed in space about 1922, the great expansions which have occurred in those social controlling factors should clearly have brot a great rise in per capita patenting, instead of none at all. The percent of men engaged in manufacturing and mechanical industries correlated +.76 with patenting, and increased to 1.36-fold in 1885-1930. Primary horsepower used in industry correlated .5, and increased 7.25-fold in 1889-1929. Per capita wealth correlated .84, and (measured in stabilized commodity dollars) rose to 1.68-fold in 1890-1922. Secondary school attendance by boys correlated .64, and absolute attendance increased 16-fold during 1890-1930. We cannot ascertain statistically from such data the increase of patenting which *should* have taken place, but very clearly it should have been large, several-fold; yet there has been none, which justifies our statement that the disposition to take patents is rapidly declining. It is strange that so demonstrable a fact has past so unnoticed.

Why Patenting Tends to Decrease

We may consider the plausible explanations for this tendency to decline, in two groups—those reasons which do, and do not presume a decline of inventiveness, which is a wholly different thing from the disposition to patent inventions. There are active fields of inventive progress, like navies, armies, astronomy, surgery, biologic agriculture, and bacteriology, in which with all their inventiveness patenting is almost zero. But first let us mention four

Explanations Postulating a Decline of Inventiveness

1. We have previously discust under Prin. 19 the tendency of the standardization which accompanies big business and government control, to discourage a few inventions which would contravene the standards.

It is not alone mechanisms like railway tracks and phonographs that are standardized and hence circumscribe new inventions, but the world's whole life is regimented, by such agencies as the movies and national advertisers setting styles of dress and furniture. It were thus possible (tho not probable) that the number of different kinds of things wanted, and the number of different situations arising that might suggest an invention, should not grow even so fast as the population. Looking at the same possibility in another way, increase of population and of standardizing forces might entail an increase in that proportion of inventing which was wasted duplication of idea.

2. The related Principle 20 which we discust indicates the growth of fixed capital, and the use of more enduring materials and satisfying, simple art design, to be anti-inventive forces, since they check the demand for new construction.

3. There is good reason to think that the native ability of the American people has been declining, thru dysgenics, and immigration latterly chiefly of the poorer classes, come mostly for purely economic reasons. It is certain that for several generations past the stupid have been breeding at a much higher net rate than those with native intelligence, and that as to acquired intelligence, i. e. education and other such advantages, socially transmissible, the distribution of child-raising has been equally perverse. The influence of such factors has been thot visible in the case of Literature, a field probably not so subject to outside influences, and allowing comparison between the living and the dead, and in which field the decline of American proved genius per capita has been noteworthy. Almost surely the same factor is to be seen in the decline of patenting, altho it cannot be proved thence.

4. As inventors with everybody else become better informed, they would waste less time inventing chimeras, such as perpetual motion machines and hopelessly complicated mechanisms. Furthermore, we have seen that nautical and presumably other patenting and invention is moving increasingly into the hands of engineers, chemists and other well trained specialists, often working for a corporation. Dr. Rossman [210,211] and Olken [139] cogently claim that engineers are trained in their schools to abhor the name of inventor, regard too highly engineering tradition and the supposed voice of authority, shun especially fundamental invention, normally out of their line, and seek to make familiar things better, not radically different. And to this modest outlook their corporation is likely to limit them explicitly. All this is not to say

that technical training and corporation laboratories are strangling invention—we would only point out that they have certain (probably more than counteracted) tendencies in that direction, opposing the otherwise expected great expansion of patenting.

Explanations Not Postulating a Decline of Inventiveness, but Relating Only to Patenting

The four foregoing explanations which postulate a decline in the disposition or ability to invent, may be accepted right alongside the 7 following explanations which make no such claim, but postulate influences tending to reduce the disposition to take patents, irrespective of whether the tendency to invent, or actual invention, were diminished or increased. At the start of our discussion of patents we pointed out that most of our Principles, and we may add, the usual, common-sense view of History, and the history of the ship, indicate that inventiveness and invention are actually increasing rapidly, the forces tending this way seeming more numerous and strong than the opposing (however valid these last be). We concluded in connexion with the life cycle and apparent decay of inventions, that there was no reason to think that invention in general does not increase.[38] Nonetheless percapita patenting has remained stationary.

5. We have explained under Prin. 5 how with advancing technology and theoretic understanding the standard for invention is raised, so that a novel combination which might have been patentable in earlier times, might today be held by a patent examiner, judge or any competent specialist as not embodying sufficient originality or difficulty of contriving. We are referring here to the effect of the increase of knowledge: in the following section we shall find the same result of unpatentability reached thru an expansion in the nature of an invention.

6. Probably inventions can no longer be sufficiently defined by patents. Their ever increasing number, with six million patents on record,[106] has forced narrower definitions and claims, so that the broader essence of the invention becomes unpatentable. The earliest patents did not describe the invention at all, but only named its effects. Present patent practice reflects a somewhat later condition, wherein the typical invention was a simple, concrete mechanism, process or combination, which might be diagramed and described in engineering terms. The new type of an important invention, the only one worth the thousands of dollars for legal struggles which an actively functioning patent often asks, seems to us to be less of an engineering and more of a commercial concept. It seems an investment of thousands of dollars at piling up detail inventions and adjustments along a certain line,[107] fighting legal battles, organizing production, winning sales, training a variegated staff as a going team—and it all begins with a concept that a certain need is worth trying to meet and could probably be met by all these kinds of effort directed along a certain general line (Prin. 2). For an invention so considered, the patent system gives quite incomplete, short-lived and hazardous protection. It becomes often easy lawfully to seize on the nub of a good idea, without infringing its patent. Thus patenting may become not only useless but detrimental to an inventor, thru advertising his ideas among

[107] Hatfield (*Inventor and his World*, our note 161, his p. 247) says that usually an invention is essentially the perception and correct formulation of a need and the commercial characteristics necessary for a successful solution, then a number of alternative solutions prepared by the same inventor. The first is unpatentable, and the second with its requirement of plural patents is burdensome, especially internationally, and hardly admits of assurance that every possible solution has been covered. Sir Wm. G. Armstrong said long ago that the chief work and value of creating an invention is the working of it out in practice, which the patent laws little helped. *Recent Discussions on the Abolition of Patents* for Inventions in the U. S., France, Germany and the Netherlands—Evidence, speeches and papers in its favor. London 1869. Armstrong's claim is cited by Dircks (N 154).

his rivals, and selling him at a high price a dangerous right to start infringement suits.

This growing insufficiency in patents has been met partly by plural patents on alternative processes, and partly by other institutions than patents, viz. the trade mark, design patent, copyright, fair trade laws, advertising, and monopoly tactics.

7. The patent system is an old institution, hardly changed in principle in the past three centuries, and surviving in its most primitive, simple form in America. Yet it is the mechanism by which we attempt to pay for and direct the most progressive and incessantly novel activity of our whole civilization, invention. It is an old bottle to hold very new wine. It is as if someone should propose to finance and control that other most progressive side of life, the Sciences, by reënacting a statute from the age of the Stuart kings. It might still work, but one would expect increasing disregard for it.

Signs of such disregard are not alone the falling behind of patenting. There stand Henry Ford, who disliked patents, Ives, and Weston, one of our best electrical inventors, who practically gave up taking patents, considering that fighting to protect them consumed too much of his time. Patents and secrecy are considered alternative defenses for inventive property, so that either might be expected to increase if the other diminished; yet both are declining. It is especially the larger, more progressive and in short more typically modern firms which abandon secrecy. Sometimes they have formed cross-licensing agreements or patent pools which are essentially an abandonment of patenting (i.e. the monopoly of an invention), except as a means of keeping outsiders outside the industry. The proposals often made of late for giving a property right in the industrial uses of scientific discoveries (disapproved by a competent committee[108]) do not envisage real patents, i.e. private ownership and monopoly of the discovery, but instead some system of freedom to use on payment of a decreed royalty, or else an award of prizes. . . . . The American Economic Foundation has moved for general patent reform.[108B]

Patent monopolies have some resemblance to the Renascence gilds, and may be suffering as the gilds did when an age came along that demanded larger production and progressive methods. The modern age, whose twin *Zeitgeist* is Science and Organization, honors engineers perhaps higher than lawyers, and demands rationality, progress, and production by great organizations combining all the best ideas known. Our age dislikes the obstructiveness of both patentee and secret process man, their annoying, blocking ways of forbidding people to use the methods that are logical, bidding them instead to go on using the antiquated means, or devise a new and probably inferior way, or beat the patent or secrecy by means fair or foul, or reach an agreement of sale or royalty thru bluff, caprice or weakness. (There can be no market price where the good sold is unique, and worth much more to the buyer than the seller loses in giving it up.) . . . It is significant that the people who most often refuse to take patents on valuable inventions are the physicians and scientists, who are those most in touch with that modern spirit. Sometimes seeing a need of monopoly control they take patents and give them to a foundation, which does not have the obstructive, self-seeking aims of small private monopolists.

But mild public disfavor would not be enough to weaken much an entrenched institution which was profitable to its patrons individually. We shall show in connexion with unwanted inventions, Prin. 37, that a thing may flourish which nobody wants for all, but some want for themselves. The whole class of nuisance-value and drag-net patents is an example. We could hardly explain the great falling off in the disposition to take patents unless patents became worth less money to the patentee. This has probably come about thru the narrower definitions of the patent office, a possible growth of legal costs, and of infringement arts, and an indubitable modern disposition of the courts to destroy patents and uphold infringers. Especially since the 80's the Supreme Court has been rejecting patents freely for lack of inventive quality.[108.1] Patentees are successful in recovering for infringement in only 10-15% of cases; even when the patent was upheld they can usually recover no compensation for past use. The patent was destroyed in 60% of Supreme Court cases, since 1880; but these conditions are not so severe in the lower courts. Here we see an important way by which that general sentiment against the obstructiveness of patents bites home to the examiner considering whether to grant a patent and to the prospective patentee.

8. The increase of *monopoly* is probably another great reason for the strong decline in the disposition to patent. To a perfect monopoly patents would be only a useless bother. True, no monopolies are perfect and sure to remain so, and most of the monopolistic concerns, like the shoe machinery trust, are taking out and buying up abundant patents, and clearly using them to fortify their monopoly, as Vaughan [180] shows and deprecates. Nonetheless it seems likely that these or other monopolistic corporations are using less patents than they would, if they were again subdivided into numerous competing firms.

9. Larger views of bigger businessmen may be another explanation of the disinclination toward patents. The normal business executive is becoming a college-trained man, in touch with the technic sciences and economics, as discust under 7 above, a gentleman, of social and coöperating disposition, and an employé and shareholder rather than the owner of a firm he built up by fierce competitive

struggle. Such a modern executive is disinclined to the narrow, dog-in-the-manger, self-serving of patenting and secrecy, and inclined to do the big, urbane and cooperative thing. The changed spirit may be seen in improved labor relations, advertising, business honesty, and most clearly in the outside and inside architecture of business premises.

10. The changed nature of inventors may be another reason for less patenting. We have seen that more and more they are insiders, working for corporations or at least a particular industry, and hence less likely to wander afield with half-baked and alluring new basic ideas. This might reduce the number of patents. Again, with everyone becoming more intelligent, the multifarious costs, extreme risks and normal loss thru ventures in patenting may have become better known, deterring from the patent office especially the free lance, outsider, amateur, and the workman and small fellow generally, who would have to look to the will o' the wisp Patents for his reward, instead of to a salary or increased profits from an established business.

At this point our original edition quoted the opinions and evidence of authorities from 1925 and earlier, that American patents were a very poor lot, rarely worth their many costs to the patentee. These judgments were probably justified at their time; but since then patents have so much improved, coincident with higher education, bigger business, and an apparent fall to 7%, during 1910–60, in the ratio of patents to presumed invention, that the author thinks best to withdraw his former strictures, and to present evidences of great improvement in the corresponding chapter of his *Supplement*. Naturally if a much smaller proportion of inventions are patented, it will be the better inventions and the tighter patents that will be retained. The recent statistical studies of B.S. Sanders (in *Idea,* vol. 1, and 2:1-60 with Rossman and Harris, and 6:87-116) provide good evidences of this, incl. that 60% of the modern American patents are worked sometime. Another evidence is that in Britain, which like most other countries requires subsequent and increasing fees to keep a patent in force, 18–20% are now kept alive to their limit, compared to 2% some years ago. The growth of governmental invention, rarely patented, faster than the commercial, has been one reason for less patenting. The matter is graphed and explained in my *Invention and the Patent System,* and in the *Supplement*—see p. ii, by the same author.

11. The large improvement which has been brot about in the American patent bar should be another reason for less patents. Many attorney sharpers who flourished 30 years ago, enticing the unwary to patent perfectly hopeless inventions, have been driven from the profession.

Conspectus of Recommendations

Before making any recommendations, it were well to give a few words to discussing the goals, toward which our proposals should tend.

One aim we should not pursue at all is to protect the natural right of an inventor in his invention. For natural rights have no existence, according to the Philosophy and Sociology of today. They are a figment of 18th-century imagination, embalmed in our sacred Declaration of Independence. Lawyers, politicians and the masses still talk about them as real; yet such concepts cannot survive five minutes of critical examination. The only inherent right of an inventor—changeable statutes, customs and contracts aside—is that it may be expedient for society—or may not be—to pay him more or less well for inventing. He has usually done something useful—and so has any miner. We show thru Prins. 1, 26, 28, 33, 34, 36 and 37, and thru the whole idea of social forces behind invention, that one inventor's contribution is neither indispensable, long unique nor necessarily worth its commercial yield.

Inventions Must be Brot into Use

Secondly, our goal is progress and welfare, not simply more inventions. We might well obtain more technic as well as social progress, with less inventions. For we have seen by Prins. 25 and 37 that very many inventions are mere wasted duplications of effort, made thru ignorance of what has been done, or is being done in the secret fastnesses of other inventors' laboratories—or thru the effort to get around someone's patent—circumvention to beat invention. And other inventions are mere ways to trick the public, to sell it what it does not want, or bigger and better wars. Again, it is not enough for progress that the invention be made and worth while—it is needful that the invention become known to the right people and be brot into use, driving out the old (altho as we have said there are many cases where an invention better than all the prior art nonetheless ought not be brot into use).[109.1]

We have been looking at some of the many and strong forces opposing the introduction of inventions already made. To help these we have many means of spreading news, such as journals and advertising, but few formal societal institutions to propagate new inventions, except in agricultural lines, where our Dept. of Agriculture does important propaganda work. In the similar matter of encouraging utilization of medical, psychological, educational, agricultural and other such discoveries, the Children's Bureau, Public Health Service etc. do something. But inventions proper, especially patented ones, our bureaus fear to talk about concretely, because commercial interests of particular firms are strongly involved. Private profits are sacred, and bristling with fighting power; the public, the consumers, may only be vouchsafed a little enlightenment, provided no particular vested interest be

109.1 Patents operate in the reverse direction, thru royalties taxing the new in favor of the old. From our consideration of the modern rudders we have had reason to think they would likely be in fuller use today, and better forms, combining more good principles, if there had been no patents to obstruct and tax the utilization in rudders of the aerodynamic ideas gained thru airplanes and turbines. . . . Cf. our discussion of Prins. 19 and 20, and Invention by Wholesale, N 201.

impugned. If we mention Russia now some people will see red; but it is merely interesting to observe that there is a country in eastern Europe where, altho the problem of vested private interests has been set aside, the administration thinks advisable to maintain a special government bureau to follow up approved inventions and see that they are actually put to work, and not left to shiver on the doorsteps of officers who may have a personal, tho not commercial vested interest in the old methods, or in indolence.[265] . . . . Let us recommend some improved means or other for encouraging the use of good inventions. Doing so is most difficult under the patent system, which entrusts this whole aim to an attempted private monopoly of each new device, but is much easier where that grip is relaxed, as by means we shall propose.

Patents Fundamentally Considered

It is said to be a fundamental principle of good inventing that faulty working should be remedied not by adding an auxiliary mechanism, but by striking at the root of the difficulty. So in our social inventing, it is time to face and deal with the patent system's fundamental defect, which is that *by attempting to make private property of a set of ideas* (an invention), a thing inherently most difficult to define, name the author of, and protect, we call forth

110 Dr. Barkev Sanders tells me that he found the highest correlation between frequency of patents on an invention, and use of that invention, when trying a lag of 17 years. This suggests, tho we should need other reasoning to prove, that the use of the invention was obstructed so long as the patents were in force. Again, I have known a case where a manufacturer was interested in an invention offered him, but on looking up the patents, and finding 120 in the same field, he dropt the invention without further inquiry as to its novelty, not wishing to enter so dangerous a field.

111 Jn. R. Gammeter, chief of the Goodrich rubber invention laboratory, has written: "There never was any one man so bright that someone else cannot invent a machine about as good as the first man invented. Why not adopt a policy of fair play, and cut out all these lawsuits? If manufacturers in similar lines would pool their patents, instead of eternally trying to sue each other in court, there would be better business and more profits for everyone." P. 196 of We Make Inventing an Every-day Job; in *System*, Feb. 1925, pp. 181ff.

multifarious and prodigal struggles [109.2] over the ownership of ideas, likewise assessments and prohibitions on their use,[110] and furthermore neglect of fundamental, unpatentable, litigious,[111] revived and other inventions such as the 17-year-pretended-monopoly system practically fails to reward.[141]

Our Recommendations

There is need of a thorogoing, basic reconsideration of the patent system and the whole problem of how the needed kinds of invention may best be fostered (and possibly hurtful kinds represt), and the use of the best inventions encouraged. The problem is too vast and many-sided to be solved by the present writer, nor, probably, by any single person. For invention is an effect and cause of all sides of life, involves all sciences and industries, and centers today in two very special industries of its own—the patent law business and the Patent Office. A study of the needed scope, thoroness, many-angled viewpoint, and authority for influencing legislation, can only be made by an accredited, competent and leaderly commission, combining all the principal competencies needed, and working for no special interest but for the country. It were best financed by one of the social research foundations with the coöperation of the government, or be created by Congress. This Commission should have prestige to obtain important witnesses and private documents, and should be crowned with some power to influence Congress; and it will need considerable funds for research along untouched lines. It should be composed of a Patent Office official, a patent lawyer (only one), a free-lance and a corporation's inventor, a representative of big business, a manufacturer, a technologic historian, a sociologist, an economist, and a political scientist expert in legislation, the last four representing the public interest as well as their very pertinent sciences. Care should also be taken to represent thru these men Chemistry, Electricity and Mechanics, and no attention should be paid to geographic, political or other extraneous considerations.

But care should be had that the commission be not dominated by men of conservative tendency, for then it would be helpless to propose radical remedies, such as the complete replacement of the patent system, should such reforms be revealed as desirable by the research. Nor should the commission be dominated by representatives, practitioners and beneficiaries of the present inventive and patent vocations; because these are sorely tied by habit and self- and group-interest to the perpetuation of their vocation, little changed. The social scientists would be the most pertinent and valuable members, on a commission considering the modification of social institutions, our political and economic arrangements to encourage the making and use of inventions. The commission would assemble to study not invention, but the social control of invention. Anyone who knows anything of social science knows that vested interests do not reform themselves. Only *outside* study, pressure and probably force can do that. As with invention by our Prin. 32, from insiders we expect minor betterments, but the fundamental reconsiderations from outsiders. It was quite to be expected that the patent lawyer chairman, Karl Fenning, of the national committee on patent legislation, representing the patent law associations of the country, should report "I think on the whole, the patent system is as nearly perfect as most things that are operated by man."[142] Similarly and equally useless for fundamental reform have been the various extensive hearings[143] before Congress, where those giving testimony were all practitioner-beneficiaries of the present arrangements, never social scientists; likewise the Report of the

Committee on Patent Law Revision of the American Bar Association,[144] and the Report of the Patent Committee of the National Research Council.[145] With a distinguished membership of three patent lawyers and six physical scientists and inventors, this Committee of course "approached its work in the belief that the American patent system has been one of the most potent factors in the development of the prosperity of the country," as they naïvely avow. Their recommendations, while good and apparently influential, were limited to four improvements in practice, nothing fundamental. No better would have worked, presumably, the commission of inquiry once called for by the Inventors Guild, of 300 of our best inventors, which proposal President Taft strongly endorsed to Congress. Will another enlightened President endorse the call for such a commission, to have a proper basis this time for fundamental consideration of a social institution? Tinkering by insiders is always highly needed; but when an institution is basicly defective and becoming increasingly out of date and abandoned, as we have shown the patent system to be, it is due for study by outsiders, possessing a bold and competent insight into institutions, or the functioning of political, cultural and economic society.

Some Problems for the Commission to Study

First some ground-laying questions the Commission might take up would be:

How many millions a week is the country spending on getting and fighting over patents, including all court and legal expenses, the time of inventors and executives while concerned with patents, and efforts at invention for circumvention.

Attempt to measure what is lost thru using inferior methods to avoid patented ones, and thru avoiding not only use but independent invention in fields covered by another's patent, or fields which might be, because thickly beset with patents.

Attempt to measure the direct and resultant costs of secrecy in invention.

A. Replacing the Patent System, Wholly or Partly

How about complete supersession of patents? The economist Vaughan concludes from his study: "The social costs of the present patent situation—industrial monopoly, suppression of patents, discouragement of invention, and waste of human and material resources—offset considerably, if not completely, any good derived from patents."[146]

Invention thru NRA trade organizations. The logical party to call for needed inventions and pay for them is usually the national organization of each industry,[111] in default of international organization. For the most direct benefit of each invention will usually fall to the industry which uses it directly, despite all the efforts of patentees to corral closer the profits, and howsoever it be true that the chief final and ultimate beneficiary is the consumers over the whole world. Hence considerable technical research and invention has been done by trade organizations in America, which as we said were spending fifteen millions annually before the Depression.[147] But then they were struggling with a fatal obstacle to an ambitious program, in that any firm which wished could stay out of the association, and yet reap most of the benefits of their research, thru the manifold failure of patents to monopolize inventions. But now, with the NRA[147.1] and the general trends toward organization, it should be readily feasible to oblige all firms to bear their share of the costs of invention and research, for the good of their industry and perhaps of still wider industrial groups, and to make the progress as intense as the majority pleases. With resolution and little legalism all existing domestic patents could be pooled, as in some industries during the war, with compensation during the next decade from the firms with the poorer strings of patents to pool, to those with the better, payment to be assest according to whether a firm *should* use the patents, not whether it *would* use them. Governmental sitting-in would protect minorities, and the public and consumer interests, harmonize the policies of separate industries,[53] and ensure respect for the general national policies, with regard to labor-saving, military power, conservation etc. Patents might perhaps be granted, without examination on certification by the trade's research authority, but they would have little significance. Foreign patents could be taken out. Private inventing might remain untouched, but the trade associations could have eminent domain over any patent to an American, and to foreigners too if treaties permit, to buy it up at a fair appraisal . . . . An affiliated organ within the trade, as in some British trade associations fostered by that government, could follow up their inventions, and others new or old, from abroad or at home, that were judged to need promotion, to see that they be understood and introduced wherever suitable. Thus turning to new ways would be encouraged, instead of taxed as by patents, nor ever prohibited. English, German [269] and American experience is available. The idea seems to me particularly suitable, in line with past developments, and a logical plan whereby for the first time the chiefly interested parties could go after, pay for, get and use the inventions which they and the nation want.

Other problems for the Commission to consider are:

How else might the charge for making inventions be spread upon the very wide classes which benefit from them, thus ensuring ample funds for exploration, by schemes which do not tax novelty with royalties?

How else might we bring most rapidly into use valuable inventions, including those unpatented in America, or with expired patents, some of which are never useful till long after their devising, when some changes in the many elements of the invention's complex (Prin. 2) at last make it valuable.

Cash awards to inventors *ex-post-facto,* somewhat as the British did for war inventors.[148]

An inventor's institute, as per the French example [99] and various suggestions [101] but greatly expanded, that would help meritorious inventors beforehand, when they most need help, and dissuade others from invention. The Inventors Foundation is now helping with instruction and advice.[101]

Can anything be done for the myriad of inventions, at least half of the total in importance, which are not effectively remunerable by patent today, and most of which are not patented?

Extending some form of reward to commercially used scientific discoveries and improved, cross-breeding strains.

Are patents needed as a reward for disclosure? Is there a real chance today to keep any inventions long secret, except some chemical ones? Might this call for special consideration of chemical invention, in framing legislation?

Should secrecy be banned by law, as today in cases of manufacturing dangerous articles, taxability, census etc.?

An immediately established government committee for the study, prediction and promotion of new or expanding industries, such as television, the telegraphone and air conditioning, especially to encourage new capital investment in an attack on the Depression. The Science Advisory Board has recommended such a Committee to the Business Advisory and Planning Council for the Dept. of Commerce, starting committees of its own.[148.1]

B. Keeping Patents, Wholly or Partly

Feasibility of variable durations [112] and scope of patents, e. g. to promote fundamental invention by allowing to a responsible great corporation like the Bell telephone company a 60-year monopoly of teletactics (feeling at a distance) by whatever means achieved, provided they worked constantly to develop this invention (for which a plan was made by a distinguished inventor 40 years ago), exploited it as soon as practical (if ever), and kept their charges reasonable. Such a concession would hardly be a patent, not being based on any particular achievement. It seems doubtful tho whether any kind of patents or concessions could effectively promote fundamental invention, since any rewards held out would be far in the future, and quite uncertain, and there are few companies with even such a long view and purse as the Bell.

Hatfield's proposals [149] of moderate patent rewards widely scattered, to include the various makers of an invention, including the rediscoverer of a forgotten idea, and the enterprizing backers.

Regulation of permissible contracts between inventor and patron, to protect the inventor and the public.[149]

How much control over the directions of patenting, between industrial fields and types of invention, (as labor-saving, etc.) is needed and feasible?

Does not the subject-matter of patents and inventions (given the social establishment of a patenting system) belong to Technology rather than to Law, and would it not be better handled exclusively by technologists, physical scientists, inventors, industrialists and bureaucrats, than by lawyers, and by judges quite ignorant of the subject-matter, even if there should be special patent courts? And would not the method of handling better

[112] Ravenshear (Note 170) proposes lengthening the British period of 14 years, for fundamental inventions. Some countries limit the life of accessory patents to that of the basic one.

be common sense and quick hearings, rather than legal formalities commonly lasting years? Stringham would mix the two thot-worlds.[108.1]

Compulsory license of insufficiently worked patents, altho this has hitherto been so enmeshed in legalism as to be almost ineffective, in the numerous countries having it. If made thoroly effective, compulsory license would profoundly change the patent system, and amount in effect to state determination of inventor's rewards, to be recovered from a tax on the several inventions.

Compulsory working in some country, not necessarily in this, and provided society would gain from the invention. Unless to attack monopoly at the cost of industrial efficiency, there would be no gain in allowing rivals to obtain use of the inferior methods a corporation has laid aside, or patented as a precaution.

Preëmption by the people of essential inventions, with compensation to the patentee, as was done by the "fossy jaw bill".[149.2]

Taxation of patents, annually increasing, as in many countries, to get rid of the futile, unworked and hold-up types.

Publication of patent applications after 2 years, to discourage delay and strike at the drag-net patent application, lying in wait for an industry to arise.

How else attack the non-constructive, merely predatory patent, without discouraging the starters of fundamental, slowly incubating basic inventions?

Since all countries make some kinds of inventions unpatentable, how about barring from patents or any other assistance all advertising devices, other means of tricking the public, and all apparently useless inventions?

Barring as patentees in scientific fields, e. g. of chemical and electrical inventions, all men who cannot qualify as competent in the science or industry involved.

The profession of inventor, and how it might be improved. Schools for inventors [159], teaching invention in engineering schools,[101,139] and better handbooks of elements for making inventions.[160]

Combining the Canadian and U. S. patent systems for research on claims; recommended by the Patent Office Society.[149.3]

All these are not matured proposals, but are samples of the sort of plans for radical improvement that should be studied, by the proposed Commission and by various other competencies than the present writer's. Above all the basic reëxamination of the patent system, and the encouragement of invention by trade associations with compulsory membership, and the immediate establishment of a New-industries Committee in the government, seem to merit attention.

CHAPTER VII

SOME EFFECTS OF INVENTION

While our study is primarily on social *causes* of invention, we should not close it without discussing some social *results* of the same. We shall make no attempt to cover this subject fully, but speak only of a few results which follow naturally from our previous study, and which have not been fully or rightly treated by other authors.

One reason why the causes and results should be considered together is that inventions, the result of social and other forces, become themselves the cause of further inventions, or of obstruction to invention. Cause and effect alternate endlessly, like the flat and vertical links of a chain, and each invention is both. Let us try to understand the endless chain, not say which link comes first. So we have already discussed under Causation some fourteen of the results of invention, to wit: They modify the *milieu*, the civilization and culture upon which further inventions must be built (Prins. 2, 6a), furnish an engineering basis, a prior art which future inventors can employ (Prin. 3a) so long as they have any applicability (Prin. 11). Inventions may have a particular significance among the causative factors because they may come into use more quickly than most other changes (Prin. 7), especially during their phase of most rapid absolute advance, (up slope of curve 4 of Graph 1, Principle 9) thus stimulating attention and perhaps snapping an accumulated tension. They call forth science as well as further invention, so that a science, like hydrodynamics, often follows tardily after inventions which should be based on it instead of preceding it (Prin. 4). Inventions that economize something will, under certain circumstances, reduce their own field, as to its size, importance, and progressiveness (Prin. 10), altho the opposing effect is more common. They usually build up the firms (Prin. 18) and regions (Prin. 16) that make and employ them, and make these the preëminent centers of further invention in the same specialties. At the same time they smite the workmen and capitalists in the altered or superseded industries (Prin. 20). They have shifted geographically the focus and the frontier of civilization, and have built up our race biologically (Prin. 1d). The perfection of invention brings beauty in the object. (Prin. 13). Great inventions and heroic inventors are sought out or cooked up to be parts of the national epic (Prin. 30).

Many Call Invention the Great Source of Social Change

The principal Principle we would inquire further into is No. 6a, the general one of inventions as causes of social change. Long slighted, and still so usually by those who interpret history through philosophy, religion, or politics, the Principle has on the other hand been overdone, I think, by those who have come to think about inventions. A Phi Beta Kappa orator [116] in 1843 presented the idea that inventions are the prime factors in history, as one that would astonish his audience. Marx popularized it, and its ablest defender today is Prof. Ogburn,[117] altho he is usually careful, especially latterly, to add scientific discovery, and not to claim the pair as the only or with certainty the strongest cause of social change. Chapin concurs. Kaempffert has been more uncompromising, and the engineers often reach absurdity in their claims that

> "We are the movers and shakers
> Of the world forever, it seems"

instead of the poets wandering by lone sea-breakers, as the original writer had it.[119] It is fortunate for social happiness and stability that so many of us can believe ourselves of the ruling class.

It is commonly seen that the new things done—the flights across continents, the broadcasted addresses, the diseases prevented—are almost all done with new inventions and discoveries. Thence we readily slip into saying that the inventions caused mankind to follow those new paths of civilization. Suppose we make a parallel observation, that almost all new things done, and old ones too, are done with *steel* instruments. Shall we then proclaim that *steel* is the mainspring (and hairspring) of civilization? That would be true in a certain sense; but yet calling this the Steel Age would leave a host of questions in our mind unsatisfied, as to *why* the steel is used as it is, the same steel pistol being used now to build up civilization and now to shoot holes thru it, the same steel ship carrying emigrants one year, cheap tourists another, and tied up a third. What shapes and moves the steel, we ask, and quite the same, what shapes and moves those *inventions?* Just as soon as we see that inventions have *causes* (and we have been finding at least 36 out of their many), then we can no longer hold inventions the *prime* cause of anything. We may indeed properly treat them for the nonce as prime, for convenience in study, not to lose ourselves in an endless backward search along the genealogy of causes. We may use inventions as a stopping or starting point; but let us not for a moment regard them as causeless and prime, nor as caused only in some inscrutable and uncontrollable manner. Just as soon as we seek a rounded, complete picture of civilization and its causation, its constant and moving forces, we at once recall the many forces back of invention, and the quantity of them we have shown in this book which are strictly social and not of an engineering or physical science character, but belong in the fields of Sociology, Economics, Law, Human Geography, History, etc. If we seek to explain some modern development, say mass foreign Tourism, we must seek not only inventive causes, such as improvements in liners and railroads, but also direct social causes, such as the spread of education and wealth, and the emancipation of women, and furthermore these and other social causes operating indirectly, thru stimulating, i. e. partly causing those same inventions. Why has the passenger ship in the last century changed and improved more than the freighter, until the biggest and most efficient ships are almost exclusively for tourists and some other passengers—why if not because some causes have increased Tourism, and so by Prins. 15, 18, and 21 brought improvements in tourist ships?[120]

Why do inventors invent ships and things? Our whole study has been an account of social forces driving them, social causes everywhere mixed with engineering ones. The real cause of social change[172] is a welter of causes, everywhere different and multiple, which passes the wit of the present writer to reduce to any general principles of priority and supreme importance. Indeed I doubt that such concepts as the priority or supremacy of anything are valid.

None the less, very respectable students of the question have thought inventions are the usually prime factors, so let us examine the reasons they commonly give.

1. That inventions come first. But do they? I know of no case in nautical history where this could be shown, that the invention came into use before the existence of a

[120] One reason for the increase of foreign travel is that it has been left on a practical free-trade basis, perhaps because the most influential class wish to do the traveling themselves, whereas the import and hence export of *goods* for all has been throttled by business leaders with a special interest in tariff protection for their industries.

social situation (engineering aside) which made it appropriate.

2. That inventions are created and diffused abroad faster than other innovations, and so by Prin. 7 are more striking and effective. Certainly some minor inventions spread fast, like the modern rudders. But the big inventions (or aggregations) certainly do not develop nor spread very fast, remembering our Prin. 8 of slow incubation, and the dragging time, 50 years or so, which we found important marine inventions take from their first plan to important use (cf. Table III). In our search of all nautical history for abrupt inventions, we could find but a dozen or so that might so qualify, and only one, the compensated compass, that seems to have won immediate adoption. Are cultural or purely social inventions slower? How old is Fascism?

3. That inventions are more cumulative (Prin. 11). This is true when invention is taken as the abstract plan, but probably not if it mean the physical contraption; and it is these concrete things that influence the mass of men, not plans used formerly or elsewhere. It may be doubted that the variety of *things* that touch our daily lives has increased any faster than the variety of cultural and non-engineering *ideas* that throng our minds, ready instantly to influence.

Counter Argument—Examining the Correlation

Perhaps the modern prevalence of the technocratic idea that invention is prime can be traced partly to the mythologic, heroic, Adam-like, then-and-there creation idea of invention, which we have attacked with our Prin. 1, but which will never be vanquished so long as the schools continue teaching it, as we found anent Fulton. By this idea we start right off with something definite and big, So-and-So's great invention, for which naturally one would expect results and not prime causes. Its only cause was the genius of So-and-So. But if one seek all the real inventors and real causes, the task becomes multifarious, and perhaps too hard for the School Board to learn and teach, even if it desired it taught. The causes of an invention ramify backward with ever-widening generations, just as the discovery's *results* ramify forward and sideward *ad infinitum,* as was strikingly shown in *Recent Social Trends.*[117] To say that America has railroads today because Stephenson invented them in 1829, (of course he didn't) is the easy thing to teach and think. It is much harder to say that we owe our railroads not only to ten million inventions and discoveries, both before and after that date, but also to such facts as that America now has a population sufficiently numerous, dense, wealthy, educated, sober, diligent, and trustworthy to pay for, staff and use great railroads. In earlier times when Americans were few and poor, part slaves, largely illiterate, and often drunk, much railroading was not to be expected, any more than it is in Venezuela today. In such countries today they possess the invention, but hardly use it. Next, if we are seeking a complete explanation of the railroad, we must go behind those social reasons, to the causes of the causes, such matters as the limitations of births, that has made possible our present prosperity, and the causes of this in turn, in scientific and inventive advances, and also in religious and philosophic change, and all that has been affecting the feminist movement. In all this boundless welter of causation, it may be that the constantly and especially clearly recurring elements of science and invention contribute a majority of the total effect; but who has even started to prove it?

With the inventive and the social aspects of life going forward together, we are at a loss to prove which was more the cause of the other. So let us find a time when one of them went backward, and see what happened to the other. The Dark Ages offer our one available case, although the author has only some impressions, needing further study. In that age, while general civilization and culture declined, it seems that almost all the previous inventions remained a part of the heritage of knowledge, at least in possession of a few people. It was the building, use, and addition of inventions that declined, not their intellectual existence. There was apparently a decline to one-quarter in the rate of new inventions, from 200 to 1200 A. D.[121] but by no means a cessation. For in this epoch the occident invented the lateen sail, fore- and sterncastles, horseshoes, modern harness, compass, rudder, gunpowder, and improved the water-clock, weapons, armor, and the steam engine . . . . So while invention was increasing, civilization declined. That does not look as if changes in civilization depended primarily upon invention. To be sure, the *use* of inventions declined, as we said. But a decline in use of invention could not be due to invention or science itself, but must be due to some other, social factors.

The Principle of Equivalent Invention (No. 34)

The best of all arguments against the primacy of invention, and a social principle of invention which we have read nowhere else, altho it is true and needed, is our 34th, that inventions are not only duplicated about the same time by identical solutions (Prin. 25), as many have noted, but also paralleled by equivalent but unlike means for reaching the same goal around the same time. Let us give some instances.

When Watt needed a guide for his piston rod, he made what he considered his most elegant invention, his "parallel motion." In a book of 1901[122] I find 20 different ways of doing the same thing. Contraception is an art known and useful for milleniums, of time and of prosperity, but especially desired by the modern world. Hence we find 18 radically different methods indicated in a recent book,[123] without counting minor variations such as using one spermatocidal chemical instead of another. Now that men all of a sudden wish to navigate the air, we find them using or trying 16 more or less independent fundamental inventions, viz. the dirigible balloon, glider-dirigible,[124] man-carrying kite, glider, airplane, helicopter, autogiro, rotating airplane,[124.1] ornithopter, undulating wing,[124.2] Flettner-rotor flier, autorotor airplane, with fixable wing, Rohrbach's paddle-wheel flying macline, rocket, auto-rotating wing as in a new kite, and the albatross's use of the wind's turbulence. All these means depend on one or more of 6 fundamental principles, alike for support, guidance and propulsion—the principles of the balloon, airfoil (airplane), parachute, rotor, rocket and spring.

In marine history we recall numerous kinds of sails, all for much the same result, the four old methods of shortening sail (brails, reef-points, bonnet and furling some), three old methods of skin fastening (sewing, clinker and carvel building), beside modern ones, the very numerous water-heavers invented as soon as men turned toward steamboats, the highly different forms of engine used afloat, including atmospheric, rotary, semi-rotary, including Ericsson's successful pendulum engine, ordinary reciprocating, turbine, hot-air engine,[125] alcohol engine, gasoline motor and producer gas and Diesel motors, and

[124] Which Solomon Andrews flew repeatedly in 1849-65. *Pop. Sci. Mo.*, Jan. '32, pp. 15ff.

[124.2] F. W. Brearey's pat. in England, 1879. A flapping wing in front caused waves to flow aft along a flexible airplane. It was suggested by the skate, who in actuating his eel-wise waves directly, suggests a new principle. O. Chanute: *Progress in Flying-machines,* pp. 143-6.

[125] Ericsson's hot air ship *Caloric* was far from being just hot air; she was practical, save for being not so good as a steamship.

storage battery; the four ways of eliminating the stoker from a powered ship—by mechanical stoking, pulverized coal, oil fuel, and motor-ship. Everywhere are alternatives possible and usually in use. There is not an element of the ship that has not been totally dispensed with sometimes, since there have been ships or other craft without a skin, keel, deck, interior, lengthiness, fish form, symmetry, normal flotation, motive power, steering, movement, anchor, and any other trait we can think of—except the powers of staying in place and tolerating wetting.

If inventions were prime causes, why should this Principle of equivalent invention prevail? Or should not the finding of one solution rather tend to inhibit the more or less useless contriving of rival solutions, at least before the institution of patent monopolies? But if, on the other hand, inventions are essentially secondary, called forth by social forces, and sometimes by scientific advance, then the observed Principle is just what we should expect, viz.: that a new configuration of forces would call forth a number of independent solutions by different inventors about the same time, some identical and others unlike, even utterly unlike, yet filling the same need. And when, as usually, the new constellation of needs and possibilities arose not suddenly, but by unceasing modification of several of its many elements, and the number of constellations at one time is indefinitely great, one would expect corresponding inventions in the helter skelter, constantly added, indefinitely multitudinous and non-definable stream, readily but not perfectly alternative to each other, which is just what we do find as the most frequent image of reality.

No Invention can Revolutionize Civilization (Prin. 35)

This Principle follows from the preceding one of Equivalent Invention, since just as the great inventor is not necessary to any invention, if there stand always more than one man ready to make it (Prins. 25, 26), so the invention itself is not necessary to any result, since different inventions with the same function appear beside it. And if a cultural revolution or any other event happen, it cannot be ascribed as solely due to that one invention, even if (what hardly ever happens) no invention but that one were actually used in reaching the result, all others remaining as background possibilities. Further to prove and explain this and the basic Principle (34), let us examine the greatest inventions in nautical history, and three others which are commonly alleged to have revolutionized civilization—gunpowder, printing and the steam engine.

Was the contriving of the first boats or rafts of revolutionary significance to civilization (or Savagery)? No, for the few interior tribes which lack them, the Pueblo Indians for example, do not differ significantly from those that navigate. To be sure, *either* a boat or a raft would have been necessary for crossing waters too wide for a swimmer with a float; but such waters can usually be traveled around; and *neither* boat nor raft was necessary, since either would substitute for the other. We are reminded once again that the definition makes the invention. Was the invention of the sea-going ship epochal, which Horace and Ovid describe with so much more of poetry than truth? Hardly, tho there has been no complete substitute for the ship—score one exception to our Principle. We possess good records of Egyptian culture during the period of the ship's invention, and observe no notable changes to be ascribed to the innovation. Was the invention of the steamboat or steamship epochal, as has been often stated? A hearty No again. The careless thinker (which means even a great thinker when in the grip of a traditional, unconfronted error) gazes from afar upon his world, sees steamships carrying over every sea an immense international commerce without which modern civilization is never imagined, and concludes that the invention of the steamer did it. And yet, even today upon the waters, the gazer from afar could see white flecks of sails, still competing. The windjammer has been so constantly and greatly improved that she could carry all the world's freight, with no great loss. We have seen that her greatest period of flourishing came after the introduction of steam; partly, indeed, because of it. Surely if steam could conceivably have failed of introduction, with all other engines that have since been devised, so that sails were called on to waft all the commerce of a growing world, passengers included, the opportunities for invention would have been so great in such a great industry, and the demands for size, speed, safety and comfort so instant, that the sailships would have come to include many types, better for fast and precious traffic than any of the clippers on which many a now elderly but still living emigrant fared to America or Australasia, and fairly comparable to steamers.[126] One type might be the windmill ship, which we have seen is older than the steamship. Oversea commerce would of course be somewhat less than today; but civilization could survive even the total disappearance of sea-borne trade, as was shown by the vigorous fight put up for four years by the blockaded Central Empires in the War. However, that such trade would not be greatly decreased by even a doubling of costs is shown by its continuance under high taxation, and by the heavy commerce today between countries on opposite sides of the world, over routes today slower and costlier than would be sailing across the North Atlantic. Let us recall that the foreign

[126] The harbor tugboat service would have been a principal difficulty, but meetable by horseboats, galleys, dock capstans and other devices, inc. smart sailing by the large crews and skilful officers needed for fast and safe voyages. Whalers, who had to, knew how to beat to windward, in a large square-rigger, thru an ice channel two ship-lengths wide.

commerce of our country carried on in the poor little ships of Colonial times was greater, in proportion to population and wealth, than today, for the same reasons that undeveloped and out-of-the-way Argentina's foreign commerce is today 2.3 times larger per capita than our own. Yet in Colonial days sailships had not even a harbor tug to help them, and were rather ill ships to boot. International trade has with the advance of civilization grown not larger but smaller, in proportion to wealth, or to domestic trade.[127]

Gunpowder is said to have blown up chivalry and the Middle Ages, when muskets pierced the armor of the knight, and cannon battered down the noble's castle. But observe that the armor was coming to be pierced at the same time by three older weapons taking on a new efficiency—the long bow of the English yeomen, the pike of the Fleming and the cross-bow of all the nations. This last weapon, so like a musket, shot a fearful arrow, bolt or "quarrel." It had been used by the Romans, but laid aside until the 10th century; it was a power in the 12th and after, against the armored men. Its equal, the long-bow,[128] is credited with having gained the battles of Crecy, Poitiers and Agincourt, the last in 1415, altho gunpowder had been invented in 1248 or earlier, and the cannon in 1313. Not only did the long-bow win victories for the commonalty long after the invention of fire-arms, but it continued in the English army even of Queen Elizabeth. The pike was used in armies as late as Charles II, and in the American navy until a generation ago, while its equivalent the bayonet survives today, and was important within the last half-century.

---

127 There are no good measures available for domestic trade; but the ratio of the American "national wealth" to our foreign trade rose 2.5-fold in 1850-1922. *Stat. Abstract of U. S.* Argentine and U. S. figures are for 1928.

128 "The invention of the Welsh long-bow is said by historians to have changed the face of European history." Montmasson: *Invention and the Unconscious,* p. xvii.

As to the noble's castle, I think a Roman army would have made short work of it, from the numbers, discipline and science of their infantry, and the high development of their fireless artillery, which could even throw huge stones long distances. The learned Bishop Wilkins in 1648 argued cogently that the Roman artillery was better than the gunpowder kind. Such non-explosive "artillerie" and "gonnes" (for those words are older than powder), had survived in slighter forms thru the Middle Ages, and could easily have been developed to any extent warranted by the growing strength of townsmen and kings, to batter to bits the feudal castle. In fine, we see that nobility and chivalry were overthrown not by gunpowder but by social forces. They were destroyed with any weapon that came to hand, once the towns and kings grew strong, and once the common soldier got enough of food, armor, intelligence, discipline, status and a stake in the country he was supposed to be fighting for, so that he would stand and use his weapons, instead of fleeing like a rabbit when the knights charged at him.

Some say *Printing* finished the revolution that gunpowder began. If by printing we mean the invention probably made by Gutenberg in Europe, but anticipated in China, properly called typography, then it is easy to find ten or twenty rival means of multiplying writing. Two of these, viz. printing by seals, dies, inked stamps and wood blocks, go back in the Occident from several centuries to millenniums before Gutenberg, but the most of them have been added in the last century or two, when they were least necessary. In addition there are methods utterly unlike the multiplication of writing, which yet serve the same function of spreading intelligence, such as the wall newspaper (from Roman times), library, museum, school, lecture, and church service in the vernacular. The degree to which all these means of enlightenment will be used or neglected evidently depends upon broad social forces, more than upon the invention of any one means.

The **steam engine** is often called the force that made the industrial revolution, by people who do not know their technologic history. Not only have many power sources been developed since the time of Watt, that have for the most part displaced the steam engine (as some would define it), but there were many in use before the steam engine obtained the least importance. There were water and windmills everywhere, navigation by wind, stream, tide and horse power on canal, river and sea, there was animal power in many forms, moving and stationary, and last there was human power, which remains most people's main reliance in America to this day.[228] Human power can become much more effective when used driving a good machine, like a big loom with flying shuttles, or a hand printing press or typewriter. Using all these power sources the Industrial Revolution was well going in England before steam amounted to much. And they could have been indefinitely expanded, and modern substitutes added, like the internal combustion and hot air engines, in the difficultly conceivable supposition that none of the highly varied ways of generating steam power was ever invented.

Substitute Inventions

It would be easy to find countless other illustrations of these Principles. There is Geophysical Prospecting, for example, a brilliant instance—half a dozen highly different means developed all at once for finding out what's underground, without sinking shaft or bore-hole, nor resting content with surface indications. Let us abandon such easy game for the unpleasant task of recording the famous inventions that come nearest to being **exceptions** to our Prin. 34, for not having any readily available substitutes that we know of, altho some might have been invented. There are the ship (if an invention), paper, which has many substitutes, but all with serious objections, compass (altho there are many other means of navigation, so that it spread but slowly), telescope and microscope (but I think ground mirrors could be entirely substituted for lenses), smelting by coke, altho we do employ several other methods, Whitney's (saw) cotton gin (but the roller gin preceded and has since largely displaced it); and the electric communication inventions, perhaps because recent, altho they all have substitutes of a sort.

There is so much glib talk about what inventions *cause*, that it were well to look into that word *cause*. For it has a perfectly definite, agreed meaning. We can say that A caused B if, and only if, B would not have happened, unless A had sooner. So when one authoritative writer, who should know better, says that the elevator and the skyscraper caused the modern city with its traffic problems, he is wrong, and the proof is right at hand, in cities of every country save America, which have quite similar congestion problems, and hardly any elevators or skyscrapers. Or again someone says that the typewriter and the linotype machine made the modern newspaper. This can be proved false with equal readiness, since Japan has a vast development of modern newspapers, using neither typewriter nor type-setting machine. If the claim had been made for the rotary press, the disproof would not be quite so easy, since all big newspapers use that. But still we should after our study of the myriad ways in which clever men adapt their inventions to changing needs, readily conclude that the inventors would have found some passable substitute, if we can imagine that the roller press had never been invented (whose essence in the cylinder seal and textile printing roller is millenniums old and twice invented, in Sumeria and Middle America). We could at least have speeded and multiplied flat-bed presses, using duplicated stereotypes.

Always our inventions are matters of partly arbitrary definition, and the reader may have wondered how much our "equivalent" inventions are really of independent origin. To be sure, there are always links, closer if the definitions be made narrower. At least one common trait, one element of the complex (Prin. 2) usually extends to all the solutions, namely a desire to accomplish somewhat the same result. Taking as an example the 15 methods of flying named, they have in common the desires to fly, to see, to be safe, etc., all need lightness, there are materials and light construction methods that can be used by some if not all, all need knowledge of the air and of aërodynamics, and most of them need motors, which may be more or less the same. And many minor parts, such as rudders, ailerons, landing gear, instruments, fabric dope, may be common to several, altho what most people call the basic principle or the essential definition of the invention, might be as different as between dirigible, helicopter, and glider. (To my mind the last and least detail is as fundamental as any "first principle"—as to engineering realities, tho not as to definitions.)

The common goal of the functional group has an additional importance in the fact that the first inventions attract attention to the goal as desirable[221] (Prin. 23), for instance that it is possible and would be splendid to navigate the air, or to guide a ship in fog, or to physically examine geologic structures from the surface, or to reduce wind resistance and generally use wind tunnels, or to put to new uses the remarkable properties of the gyroscope. In the latter cases we have changed over from an intrinsic goal, to the goal of putting to any use a new idea.

In fine, our Principle 34 that inventions are paralleled by unlike equivalents must be considerably abated by this principle that the rival inventions, tho "unlike", are not entirely so, not fully independent, but approach identity (Prin. 25) with all different degrees of nearness. But enough remains of the Principle of equivalent invention to make it a very powerful argument against the primacy of any invention in changes of civilization.

So we should beware of speaking of the consequences of a single invention. We should say, for instance, that the terrible trade of marine stoker is being abolished not by the one invention of the mechanical stoker, but by a functional group of 4 inventions simultaneously—that and pulverized and oil fuel and the motorship. And another effect of the same inventions is inevitably merged with a far wider functional group, to wit all the inventions on land and sea that cheapen or promote the transport of goods. For a world of inventions must be used to get one leaf of tobacco from producer to consumer. We should speak of the social effects of such wide and overlapping functional groups as all inventions promoting American foreign commerce, all lighting inventions, all inventions for point-to-point and for broadcast communication, for rapid, for pictorial communication.[129] But for smaller groups of inventions we may find peculiar effects to speak of, as pulverized coal, alone among the power sources, produces an ashy smoke.

Wider Definitions Bring More Definite Effects

The wider be the group of inventions considered, the surer are its social effects, the less subject to our Princi-

[129] A striking illustration of the fact that the social effects hang upon the functional group, not on the single invention, is afforded by the author's article "The Future Home Theater," published in 1912 (*Indep.* 73: 886-91) and quite correctly predicting the broadcasting achievements of the radio today and radio television of tomorrow, while yet totally overlooking radio. I was ill informed to overlook it in 1912; and one might think there were no substitutes for radio; yet still I got it right, on the basis of other inventions.

ple of replacement by alternative invention, whether one be describing the present or the past, or predicting future effects of new or future inventions. Hence it will rarely be proper to use any narrow definition when speaking of the social effects of an invention, unless it be made clear that one is naming only a tendency inherent in the invention, a direction in which it inclines life to flow, and not naming a result which that invention, by itself and without consideration of forces behind it, was sufficient to bring about. One may easily describe the *uses* of any invention, what it has tended to evoke, but not usually its *results*.

When we expand our categories of invention to the widest possible, to cover all invention and discovery, globally, the import of invention becomes greater, but still unmeasured. We have now got in our net every possible invention and discovery, of whatever industry or description, so that one can no longer rebut a claim made for the influence of some invention by saying that other inventions contributed to the same result. And we can now turn the tables and say that the great social factors such as democracy, and the growths of education, wealth, fixed capital, health, are themselves due to invention and discovery—in part. But how great a part? The answer is still a mystery. The inventive and the social factors both operate, and each helps cause the other, but how much?

It seems possible that the share of invention[130] in causing social evolution may be as small as the share it actually receives in the distribution of the nation's income, viz. around one-half the amount spent on invention. (The other half, more or less, is spent on quarrels and monopolizing efforts anent ownership of inventions).

[130] Brinkmann concurs with us in depreciating this share of invention, and aptly says that social revolutions commonly precede industrial ones. *Encyc. of the Soc. Sciences,* article Invention.

That half of the distributive share of the invention industry, would be compared with the shares the world assigns to pay for discoveries in all sciences, to innovations in the fine arts, social affairs, religion, etc. and to propaganda to spread the same, and to the small part of commercial advertising which constitutes information regarding novelties. If we thus valued invention at what the world actually spends on it, we should be following the principle by which economists usually judge values. But there is no warrant for thus assuming that the world intelligently values invention, especially novel fundamental invention. On the one hand the inventor and his backers can never retain for themselves the whole value of a used invention, as ordinarily calculated. On the other hand, consideration of the Social Principles of invention, especially our Nos. 3, 6, 23b-26, 28, 34, 37, shows that the reward of the inventor might be greatly excessive thru his simply monopolizing the future of an invention that was bound to come about then anyway. It is a fact of which patent sharpers like Selden often avail themselves, nimble fellows who jump into a new development by getting a halfbaked patent, whose only value is the chance of holding up an industry; and by the firms which buy up patents wholesale not intending to use them, but to fortify their monopoly position.

If all inventing were totally stopt for a while, as some have proposed rather fancifully, it seems likely that we should hardly know the difference for twenty years. Civilization would go on expanding, bigger and better equipment and new consumers' pleasures being added unto us simply through uninventive adaptation to such social changes as the growth of population, wealth, and education and thru the wider diffusion of inventions already made. But in time there would come stagnation and growing distress, thru failure to adapt inventively to the changes occurring, such as the exhaustion of the timber supply or the importation of new pests. In the course of centuries an utter failure to invent might destroy all civilization: but that is not to claim that invention has any particular value today, when it exists in a certain unknown, relative abundance. Civilization would also perish if all water were removed, yet water has scarce any economic value because the supply is so ample. So it might be with invention, altho a higher present value seems likely.

The Uselessness of Premature Invention (Prin. 36)

From the Principle 23a that inventions are brot forth by social need, one would naturally expect, and find usually justified, our 36th Principle that an invention which for any reason *did* come before its time remains useless and undeveloped until its proper day dawns. We have been able to observe this in marine history in cases of the steamboat, water-jet, screws, internal combustion engine, motorboat, and stuffing bubbles beneath the ship, all the cases in which we can say with any confidence that the invention or attempt was premature. This Principle would naturally follow from 8 to 15, and the rarity of the occurrence from Prin. 23, that inventors are led by perception of need and easy possibility, accidental invention being rare. But in contravention of that Principle 23 there is a point Mr. Kaempffert has made to me, that sometimes inventions make their own uses. Facsimile telegraphy and trans-Atlantic telephony have been introduced by the Bell Company without any clear idea of what they would be much used for, but trusting inventions with such unique powers would find some uses. The oversea telephone became used most for arbitrage dealings in foreign exchange. Those who invented the good cross-sections for screw propellers probably had no idea that their shapes would later be used in air screws, steam turbine blades and airplane wings, when the day should come for such inventions. Often a new scientific principle like that of radio, or a pioneer invention like the rotorship, will set many people seeking for new ways to utilize it—and some to get a patent which might have a nuisance value against whoever *might* utilize it.

Lag and Tension

Prof. Ogburn has made emphatic and popular the idea of "cultural lag," or society's delay in adapting itself to a new force, usually an invention.[169] We should understand *lag* as an utterly variable time interval between a change in one factor of a situation and a certain adjustment to it which is considered appropriate by the people of some period.[130.1] For example, the coming of the water- (later steam) powered factory was first adjusted to by a system of child slavery (called apprenticeship) which was considered a rational and good adjustment by most thinking people of the time. Later, with advancing wealth and industry and later public schools, another adjustment, the child labor law, was devised, installed, and approved. Still later yet other adjustments such as the Northrop automatic loom were found better; and the future doubtless holds in store still other adjustments.

The word "tension" used by Hart [131] to describe peoples' state of mind during a lag needs to be kept in the definition he first gives for it but later forgets, viz. an emotional consciousness of the possibility of a better way out of some situation. A situation does not put itself in tension—the tension is read into it by one who sees there a cultural lag. Another man looking at the same facts feels no tension, and explains things quite differently. Tension is a matter of each man's feelings. If we seek an objective, permanent, and scientific criterion for what situations embody lag and so *ought* to produce tension, we shall be disappointed. We can never be sure that a situation is without lag, adjusted; because we can never be sure that an immediate better adjustment could not be devised.

Inventions that Come Unwanted (Prin. 37)

We have doubtless given the impression, although unwillingly, that all inventions are answers to social need. This is true only of the large majority that reach importance, but many, and likely a majority of all patented, are quite the opposite—mere nuisances which come because some individual or corporation or army or navy think it will profit themselves, however hurtful to society. The "need" referred to in our 8th and 9th Principles is an individual, not social, need, and is an inner aspect, as MacIver calls it,[132] which unites with the outer aspects to make up the situation of Principle 27. Among the 9/10 of the patented inventions which according to an estimate in Vaughan[133] never pay for their legal and other fees, there must be many such, if only because of failure. Patented inventions are often worse ways of doing something, which people who may or may not own the patent on the best way, take out in order to get an advantage over other producers, forcing them to use some still worse way, or not produce, or come to terms. The amount of invention which is simply circumvention might be guest at from the four million copies of patents sold annually from the U. S. office—altho of course these serve useful purposes too. Another class of hurtful inventions are the adulterants and miscellaneous ways of imposing on the public. American patented inventions are supposed to be useful; but this means only mechanical or theoretical operativeness, even tho crude or impractical, and the provision is little enforced. The only class of inventions specifically barred from patent is that called "immoral" articles, largely the supremely desirable better contraceptives. Veblen[134] points out that an invention which was never desired by anyone but its sellers may thru competition be imposed on all. Who wants the universal adulteration of Chinese tung oil, or the Maxim silencer, or red fingernail paint, or every kind of advertising device? I myself worked long trying to invent a neon sign, not that it would do a bit of net good—it would chiefly use up wealth to make night and day gaudier and more mendacious—but only that it should send money and reputation my way instead of to someone else for inventing the same thing or perhaps worse. A good change in the patent law would be to recognize the uselessness of by barring patents for, all advertising devices, as well as any other means that could be legally defined as foisting upon the public what it does not want. Another large class of inventions is the military ones. These are desired, of course, by many officers or even by a nation generally, to be able to inflict more death and destruction on its rivals at the cost of more millions; but mankind as a whole reaps no benefits, probably, except from the few military inventions which are useful for defense and not for offense.

Any kind of hurtful invention, military, advertising and what not, may have some useful incidental, or unexpected results, as wooden ship building was largely developed through men-o'-war, but there is no indication that such benefits go far toward paying the colossal costs of the merely competitive industries, including advertising, crime, and war. Suppose something *was* learned about welded ships from the German pocket battleships—the same things have long been learnable in the construction of useful ships, nowadays when there are great peaceable ship yards and shipping companies, with men and facilities for invention quite comparable to the naval.

An Economic Classification of the Effects of Invention (Prin. 38)

Our last Principle proposes seven categories for inventions according to their predominant economic effects. The effects ramify so endlessly by indirection that ultimately any invention is likely to enter all seven categories; but still it may fall predominantly in one, especially as to its more immediate effects. Most people think of inventions only as saving labor, or producing a new good like radio; but there are also inventions that save capital instead, like multiplex telegraphy, fire protection, and especially for the future most lighting inventions, since they encourage using buildings and other capital for more than one daylight shift, as has always been done with ships. Then there are the land-saving inventions like skyscrapers, fertilizers, and geophysical prospecting. And there are those which save nothing, but provide us some new or far more abundant consumption good, like the radio and television; and there are minor categories. Taking a list of 119 inventions made or rapidly crescent in the last generation and having greatest significance for civilization, which I had prepared for *Recent Social Trends*[237], I revised a former repartition[246,231] to include 8 categories, with the following results for which no accuracy can be claimed. The figures in parentheses refer to another list of the 63 most promising inventions (for social effect) of 1932, and all but the very last are percentages of the total classifiable: Labor-saving 40% (16), capital-saving 15% (27), land-saving 9 (13), saving life or limb 6, new consumption goods 24, military 4, private profit and public hurt 1;[134.5] and unclassifiable 24% of the (greater)

[134.5] The small showing of the hurtful class is probably because our counts were only of important inventions. The much patenting for forestalling purposes, and the new ways of tricking the public discust under Prin. 37, which must be little known to be effective, would hardly appear in such lists.

total. Comparing the first three categories, which are about all that affect the problem of labor's share in distribution of the national income, it appears that around 2/5 of the inventions of the last generation have tended to raise the relative share of labor and 3/5 to lower it, while among the inventions of 1932 the proportions were more advantageous to labor, 7/10 and 3/10 respectively. What should matter most to labor is not its relative but its absolute share, and almost *all* the inventions tend to raise that, except the military and private profit categories, 5%.

As Dalton[135] points out, an invention is either complementary to, i.e. elicits, or else rival to each factor of production, being rival to any one for which it reduces the relative demand. Toward industry as a whole every invention must be complementary [unless it be one of the undesired contingent], but toward labor it is sometimes rival, and very often temporarily rival to a particular kind of labor, unless and until this can change its trade, or is rival to some kind of land or capital narrowly delimited. What saves labor from being often hurt is that the typical labor-saving invention increases the production or transportation of a staple, standard manufacture or raw product, above all of factory goods; and these are mostly bought by the pay-earning class; so that what the worker loses from his pay envelope he more than regains at the store. Hence the numerous inventions we have just been counting, as hurting labor *relatively,* would be mostly turned into benefits absolutely, as Profs. Paul H. Douglas and Director have most lucidly explained.[136]

Pigou [137] points out that the simple principle we have been propounding, that an invention which economizes one factor more than others hurts that factor relatively, is uniformly valid only for immediate effects, and only uniformly where the share of that factor in the production is the average share. The normal share of labor in production (and distribution) is 60-65%.[231] For example, in an industry where the costs (or share) of capital and land were but a quarter of the normal, an invention which saved half the capital and a third of the labor would function as a labor-saving invention. For the labor displaced into other industries would make a larger proportionate increase there than would the evicted capital, and in the long run the wage in any industry is set by the general and marginal industries, set universally and not in any one. But dogmatism is difficult, because the long-run ramifications may turn up all sorts of influences; e.g. an invention like the automobile that encourages the well-to-do to spend, will tend to reduce the supply of capital and so raise interest rates and depress wages.

All in all Prof. Douglas[231] finds that his "production equation accounts for the product in terms of the growth of labor and capital without necessitating any recourses to the course of mechanical progress and invention." In American manufacturing in 1899-1922 he has shown [138] that capital and land measured in stabilized money increased 331%, or 7% annually, compounded, while wage earners increased 61%, with labor receiving 74% of the product, steadily till 1922, but not since. However, the high correlations (.93, etc.) obtained by Douglas ignoring invention and treating the production of manufactured wealth as a function of the proportion of capital and labor, are no higher than those obtained by Copeland[234] considering the production as a function of labor and *time* (i. e. inventions and other improved methods), the most efficient combination of labor and capital being assumed as determined by the existing state of the arts. So the whole question needs more investigating, as Douglas says.

An economic effect from the direction of invention, which may be quite important, has been mentioned citing Rossman [210,211] and Olken.[139] They say that the training of engineers is in the direction of analysis rather than synthesis, gives them an exaggerated respect for precedent and supposed physical laws, pride in their profession and knowledge, and abhorrence of the name of inventor. Hence altho they are bound to do much inventing, they are inclined to avoid the revolutionary (Prin. 32), and anything outside the course of their business, and to take pleasure and pride in just building things bigger and better than ever. The natural result, says Olken, would be too much rebuilding of plant, an overproduction of capital goods, which is just what many say[140] led largely to the crash of 1929, and to a deficiency of new industries and novel consumer's goods, which are just what might have obviated the former. As a remedy Olken would put invention into the engineering curriculums.

We do need more invention, above all new fundamental inventions of the sort that can not support itself commercially—that seems clear. We have pointed out that the importance of invention has been much overrated by the class having most to do with it, and that no one has ever proved what its real value is, the share it has in originating social change. But still it is clear that invention is very valuable (especially the working up tho not the proposal of fundamental inventions), and that we should arrange our patent, bureaucratic, philanthropic, commercial and educational systems so as to obtain a

140 But the authors of a statistical study of *America's Capacity to Produce* claim there was no increase in plant capacity in the manufacturing and mineral industries in 1900-29, judging the constant 20% excess capacity by "the general pattern of operative and commercial organization then prevailing." If this be true, so that the fault we are naming did nothing to bring on the Depression, still the absence of that fault might have mitigated the Depression. Book by E. G. Nourse & associates, Brookings Inst. pub., 1934, 618 pp.; esp. p. 421.

much faster flood of it, especially in the most needed directions. And that the first best step toward this end would be a commission for a thoro and general inquiry, representing wider competencies and interests than just the patent industry. And we likewise need to do some social planning, not only to guide the inventions, but also to meet their often hurt-bearing impact.[237]

⚓

Our ship has come to harbor. Now to utilize the cargo. *Can* it be used, for further upbuilding of the sociology of invention, and first and last for the *better ordering* of our societal life?

What use can you put it to?

CITATIONAL NOTES

When two numbers are separated by a colon, like 7:122, the former is a volume number.

4. For worse howlers, see my article "Who Invented It?" (N305); N140 of *Inventing the Ship;* and the corresponding text in Chap. II of that book, anent the process of myth-making on inventors.

5. F. M. Feldhaus: *die Säge, ein Rückblick auf 4 Jahrtausende.* 1921.

6. Cf. also Hart (N159), pp. 511-17.

7. *Mariner's Mirror,* 12: 107. Cf. also Jos. Löwy: *Was sind und wie entstehen Erfindungen?* 1907.

8. Ellsworth Huntington: Climate and the evolution of Civilization. Chap. 11 of *The Evolution of Earth and Man,* ed. by G. A. Baitsell, 1929. Pp. 364-78 well expound this theory, and mine of coldward progress due to invention. N 291.

10. Robt. M. MacIver: *Society, its Structure and Changes.* Ray Long & R. R. Smith pub., N. Y. 1931. Esp. chap. 26: Social Causation. P. 402 quoted.

12. F. A. Magoun: *The Frigate Constitution,* gives a list of developments in the sailship, pp. 145-7.

15. *Inventing the Ship,* Fig. 41.

16. A. H. Walker: *Patents.* 2 vol.

17. Justice Brown in 141 U. S. 427, 1891.

22. Chapin (N 152), pp. 334-6 and 282. Selden got U. S. pat. 549,160, kept secret and pending for 16 years while the industry grew fat for slaughter.

23. *Ibidem,* pp. 353-5.

24. Midsummer Night's Dream, Act. IV, 219.

25. King Henry V, Prolog, 12, and Act IV, Prolog, 49.

27. *Recent Social Trends* (N 237), p. 1:131 and thereafter in connexion with the several inventions.

28. Undersökning N 221, esp. p. 58, with stat. on beverage patents.

29. *Idem,* p. 36.

30. Kuznets N 166, p. 30.

31. Quoted in the same, p. 11, from Jul. Wolf; *die Volkswirtschaft der Gegenwart u. der Zukunft,* Lpz. 1912, pp. 236, 7.

32. See discussion at end of Chap. VI, and Gilfillan N 246, pp. 836, 7.

33. *Power,* Mch. 19, 1929, pp. 490, 1.

35. Tarde N 176, p. 180; and Chapin N 152, pp. 360, 1.

37. B. Weinberg: Les Lois de l'évolution des découvertes; in *Rev. Gén. des sci. pures et appliquées,* 37:43-7, Jan. 31, 1926.

38. Hart (N 159), chap. 4: Indexes of Technological Mastery, or Tech. Progress.

39. See our Select Bibliography on the Psychology of Invention.

40. MacIver N 10, p. 531 etc.

45. *Cultural Change,* N 152, pp. 352 etc.

46. Tisell N 221, around his p. 54 cites numerous marine and other instances of the importance of Prin. 7.

49. *Lefebvre* des Noëttes, R.: *L'Attelage, le cheval de selle, à travers les ages.* Paris, A. Picard pub., 1931. 2 v., 312 pp., 500 il. The most perfect book we know in technic history; a rev. ed. of his *La Force motrice animale.*

52. Warner N 275, p. 316.

53. W. M. Grosvenor: The Seeds of Progress; in *Chem. Mkts.* 24:23-6, 1929. Discust to our mind in *Sci. Am.* 141:267, 8.
F. B. Jewett, V-P. of Am. Tel. & Tel., says the laboratories menace other industries, and need coördination—Modern Research Organizations and the Am. Pat. System; in *Mech. Engg.* 54:394-8 & 450.

54. Research, Science and Invention, N 270, pp. 332, 4.

55. Kaempffert, N 201. Also Handicaps of Quantity Production; in *Lit. Dig.,* Mch. 17, 1928, pp. 71, 2.

58. *Psychology of the Inventor,* N 171, chap. 7. P. 118 quoted.

59. *Hist. of Mech. Inv.,* N 179, p. 19.

61. A good account is in Usher N 179, p. 286 ff.

62. E. K. Chatterton: *Sailing Ships,* pp. 9-18, discusses the conservatism of the ship; and Hobson N 163 the detail-accumulation trait of invention.

64. Chapin N 152, p. 253.

65. *The Story of the Rotor,* 1926, tr. and pub. by F. O. Willhofft, from *Mein Weg zum Rotor.* P. 10.

66. L. L. Bernard: Invention and Social Progress; in *Am. Jol. of Sociology* 29: 1-33, 1923. Likewise Chapin N 152, pp. 353, 4.

70. Rubái'yát of Omar Khayyám, Fitzgerald tr., 2d ed., 67th quatrain.

71. Sadi in Gulistan.

72. Wm. F. Ogburn: The Great Man vs. Social Forces; in *Social Forces* 5: 225-31, 1926, and the same author's *Social Change* (N 169), Pt. 2, Chap. 5, Inventions, Mental Ability and Culture; with Dr. Dorothy Thomas. The same list is in "Are Inventions Inevitable?", in *Pol. Sci. Q.,* Mch. 1922, pp. 83-98; and in Rossman (N 171), p. 142. An excellent study containing a list of about 40 inventions made or claimed to have been made independently by 2 or more men, and 108 such discoveries not inventions. Primitive duplicate invention is discust by H. S. Harrison: Evolution in Material Culture; in *Nature* 126: 726-9.

73. N 171, pp. 131-3.

74. *Social Factors in Medical Progress,* N 174, Part II, chap. 3, with long list . . . Fessenden (N 90, pp. 123, 4) denied that real inventions are duplicated; but he denied too many things.

76. The data for 1890-1930 are from the *Biennial Survey of Education.* (The number for the last year was 46,720 for "certain courses in engineering.") Prior data were obtained by near interpolation from the graduates in engineering of 37 colleges and U's, fitted to the later data by means of the 1890 proportions; from B. B. Burritt: *Professional Distribution of College and U. Graduates.* Bureau of Educ. Bull. 19 of 1912, p. 143.

77. On Mech. Invention in its Relation to the Improvement of Nav. Arch.; in Inst. of Nav. Arch. *Trans.*, 1:145-59, 1860. Cf. our later stat. on the occupations of marine inventors (Tables I & II).

78. On the Paddle-wheel and the Screw-propeller, from earliest times; in Soc. of Arts *Jol.*, 1858, 6: 335-42, esp. 335.

83. Tisell N 221, pp. 102, 3 and Figs. 17, 18.

88. Wm. C. Church: *Life of Jn. Ericsson;* N. Y. 1890; 2: 241.

90. Reginald A. Fessenden: *The Deluged Civilization of the Caucasus Isthmus.* Boston, T. J. Russell Print, 1923. Pp. 99-128 and 132-6 contain a vigorous and well documented discussion of modern inventions, especially in electrical fields as carried on by government boards, and corporations, and by the highly successful author.

94. N 237, 1: 163-6 and note 12 on p. 132.

98. By Ralph Rabbards: A Coppie of Notes delivered to her Majestie . . . . Queen Elizabeth. Reprinted by Dircks, N 154, pp. 196-202.

99. *Recherches et Inventions,* revue mensuelle, is their organ, J. L. Breton, Dir. The June 1932 issue contains a book-long article on the work and history of the organization .

100. N 237, pp. 163, 4.

101. Vaughan N 180. C. Roy Watson: A Foundation Proposed: in *Pat. Office So. J.*, 12: 387-90

Fessenden N 90, pp. 135, 6. The Inventors Foundation, 122 E. 42d St., N. Y., is described in *Pat. Off. Soc. J.* 16:589-90 & 17:84-6. Regular classes, correspondence course and individual advice are given, and the Foundation has published *Inventing and Patenting,* by Howard Wilcox and Alf K. Berle, 1934.

101.5. See E. Teale: Planes that go Straight Up Open New Fields for aviation; in *Pop. Sci. Mo.*, 126: Mch. 1935, 13-15 & 116.

102. E. H. Pierce: A Colossal Experiment in "Just Intonation": in *Mus. Quarterly,* 10: 326-32, 1924.

103. *Sci. Am.* 113: 138ff., 1915. F. C. Brown and Fournier d'Albe, 2 inventors.

104. Hatfield (N 161) in his chap. X tells of this and other fundamentally new power sources.

104.1. *Technique of Social Progress,* N 159, pp. 620-32.

105. See ftN [43] and N 194.

106. 6,192,873 at the end of 1931, says the chief clerk of the Pat. Office. These largely duplicate each other; but most inventions were never patented.

108. The Protection by Patents of Scientific Discoveries, N. Y. Science Press, 1934.

108B. The American Economic Foundation, "organized to secure for the American people the economic benefits of science and invention in industry, by opposing the abuses of patent monopolies, patent pools and royalty dictatorship, in restraint of trade", Percival White, Dir., 20 Exchange Pl., N. Y. Dr. Warren M. Person, consulting economist. Other objects of attack are said to be excessive litigation, and suppression of inventions.

108.1. Emerson Stringham: Techniker und Juristen; in *Pat. Off. Soc. J.* 16:523-4; replies in 16:745,6 and 827-31. Willis B. Rice: A Court of Pat. Appeals, 17:18-28, esp. 20-1.

109. Edw. Thomas: Are Patents Worth While? ; in *Pat. Off. Soc. Journal* 13: 232-5.

109.2. "We heard little of airplanes for five years after the Wrights flew," because Orville Wright said "We decided we would be absolutely lost if our patent became known before we had $200,000 to fight with. Our experience in the courts has indicated that we did not over-estimate the money needed." From a good article by W. Kaempffert on Suppression and Compulsory Licenses; in *Outlook* 106:837,8.

116. Job Durfee: Influence of Scientific Discovery and Invention on Social and Political Progress; in *Representative Phi Beta Kappa Orations,* ed. by C. S. Northup et al., 1915; pp. 43-75. Del. at Brown U.

117. Wm. F. Ogburn: *Social Change,* Part IV, esp. pp. 268-80; and with the assistance of S. C. Gilfillan, The Influence of Invention and Discovery, Chap. 3 of *Recent Social Trends,* v. 1, 1933, esp. pp. 122-5.

118. F. S. Chapin: *Cultural Change,* N152, esp. pp. 596-601.

119. A. W. E. O'Shaughnessy: Ode.

121. From graphing by centuries the count of items, both discoveries and inventions, in L. Darmstaedter: *Handbuch zur Geschichte der Naturwissenschaften u. der Technik,* in chron. Darstellung; 2d ed., 1908.

P. A. Sorokin and R. K. Merton: The Course of Arabian Intellectual Development, 700-1300 A. D.—a study in method (of measuring sci. advance). *Isis* 22:516-24.

122. G. D. Hiscox: *Mechanical Movements,* p. 82 etc.

123. M. Sanger & Stone: *The Practice of Contraception,* 1931. Proc. of an Internat. Cong.

124.1. Curtiss—*Sci. Am.* 138: 161, 1928.

124.3. Experiments of H. W. Bull are described in *Pop. Sci.,* Jan. 1935, p. 27.

130.1. Jas. W. Woodard: Critical Notes on the Culture Lag Concept; in *Soc. Forces,* 12:388-98.

131. *Tech. of Soc. Prog.,* N 159, pp. 613ff.

132. *Society, Its Structure and Changes* (N10), p. 531.

133. *Econ. of our Pat. Sys.* (N180), p. 169.

134. T. Veblen: *Instinct of Workmanship,* p. 314; cited by Vaughan, p. 15.

135. *Some Aspects,* N 229, Chap. 7, esp. pp. 210-7. Cf. also Douglas, N 231.

136. P. H. Douglas & A. Director: The Problem of Unemployment Macmillan, 1931. Pp. 121-64, on the Fear of Technologic Unemployment. See esp. pp. 121-4.

137. A. C. Pigou: *Wealth and Welfare.* London, 1912. pp. 716-24.

138. Douglas: *Real Wages in the U. S.* 1930. p. 567. Conclusions repeated in his *Theory of Wages* (N 231), pp. 487 ff.

139. H. Olken: Invention—a Coming Profession in *Sci. Am.* 148:28-30, 1933.

141. Vaughan: *Econ. of our Pat. Sys.,* p. 217.

142. House of Rep. Hearing on Gen. Revision of the patent laws, Jan. 1912, p. 7.

143. The same; and Arguments before the Com'ee on Pats. of the House of Rep., Feb. & Mch. 1878.

Rept. of the Com'r of Pats., 1900.

Oldfield Hearings before the House Com'ee on Pats., 62d Cong., 2d session, Aug. 8, 1912; discussing largely fixed resale prices and patent suppression. Report of the Investigation of the U. S. Pat. Office made by the President's Commission on Economy and Efficiency, Dec. 1912. House doc. 1110, 62d Cong., w. appendices by A. du Bois-Reymond and Osterrieth comparing the patent systems of Germany, England and the U. S.

Hearings before the Com'ee on Pats, House of Rep., 66th Cong., 1919, Jn. I. Nolan, chmn.

House Com'ee on Pats., Gen'l revision and amendment hearing, 1932.

144. In *Reports* of the Am. Bar Asn., v. 49, 1924.

145. Reprint and Circular Series of the Nat. Research Council, No. 1. Pub. without the proposed bills in *Chem. & Met. Eng.* 20: 150,1, 1919.

146. *Econ. of the Pat. Sys.*, p. 249.

147. See our consideration of Prin. 17d in Chap. III, above ftN *50*; & N 270.

147.1. V. I. Richard: The Patent Profession and the New Deal; in *Pat. Off. Soc. J.* 16:595-9 & 678-87, 1934; good also as a general criticism of the patent system. Cf. also Forum on NRA and Patents; in *do.*, 16:657-9.

148. "Awards to Inventors." Four reports of the Royal Comn. on Awards to Inventors, 1922-6 etc.

148.1. By May, 1935, the Science Advisory Board Com'ee on the Pat. System Related to New Industries had sent in to the Sec. of Commerce the report of its sub-com'ee A on the patent system, while sub-com'ees B, on the place of government, trade asns., Univs. and private industries in the organization and support of industrial research; and C, a com'ee of indus. executives to study the general question of the relation of inv. and research to the development of new industries, were as yet unorganized. Mr. Henry P. Kendall, Chmn. of the Bus. Advis. and Planning Council, has agreed to appoint C when the previous studies are far enough advanced. Info. from Mr. Maurice Holland.

149. H. S. Hatfield: *Automaton,* N 160; *The Inventor and his World,* N 161; The Encouragement of Invention, N 197.

149.2. Vaughan proposes; p. 247 of *Econ. of the Pat. Sys.*

149.3. See *Pat. Off. Soc. J.*, v. 1, inc. Jas. H. Lightfoot: A Proposed Dept. of Inv. and Discovery, 1: 116.

BIBLIOGRAPHIES

The following Bibliographies are intended to present the more generally useful books and articles in their several fields. But none is even partially complete without adding the works cross-referenced by Note number at its head. And many good titles are to be found among the preceding Notes, both the collected Citational and the footnotes in the text. In general the preceding Notes are intended to give the most full and authoritative source on the particular point of reference, or the source of a quotation, or an authority for a statement liable to be questioned, while the works listed in our Bibliographies are usually those of more general utility. Bibliographies on Primitive Invention, and Histories of the Ship, are contained in our companion volume.

BIBLIOGRAPHY ON THE SOCIAL CAUSATION OF INVENTION

The *Journal of the Patent Office Soc.*, ed. by Dr Jos. Rossman of the U. S. Pat. Office, gives much attention to this field. Monthly, $2.50 per year.

BIBLIOGRAPHIES on the social aspects of invention are to be found in Rossman (N 171 in the following), Brinkmann N 184, Chase N 228, Hatfield N 161, Mumford N 234, Ogburn N 237.1, and Severance N 219.

Books

150. Bavink, Bernhard: *The Natural Sciences*—an introduction to the scientific philosophy of today. Tr. fr. 4th German ed., by H. S. Hatfield. N. Y., Century, 1932. Chap. 5, pp. 562-74, on the Philosophy of Technology.

151. Beckmann, Lothar: *Erfinderbeteiligung;* Versuch einer Systematik der Methoden der Erfinderbezahlung unter besonderer Berücksichtigung der chemischen Industrie. Berlin, 1927.

152. Chapin, F. Stuart: *Cultural Change.* Century pub., 1928. $3.50, 448 pp. An extensive sociological discussion, using Kuznets' data, *inter al.*

153. Dietzel, Heinrich: *Technischer Fortschritt und Freiheit der Wirtschaft.* 1922. 62 pp. A document argument against Socialism.

154. Dircks, Hen.: *Inventors and Inventions.* Engl., 1867. A defense of the patent system, by an historicly minded engineer.

155. DuBois-Reymond, A.: *Erfindung und Erfinder.* Berlin, 1906. 291 pp. Partly philosophical, w. biol. analogies.

156 Guyot, Yves: *L'Inventeur.* Paris, 1867. 479 pp. Quite out of date, but the first book published on invention from a social science point of view, in this case Economic.

157. Haeghen, Georges vander: *Le Type de l'Inventeur dans la littérature.* Liège, 1920; 71 pp. Rev'd in Pat. Off. Soc. J. 15: 488.

158. Haldane, J. B. S.: *Daedalus,* or Science and the Future. Today and Tomorrow ser., Dutton, 1924. A lively, brilliant discussion of future invention, and the "immorality" of all great biologic inventions.

159. Hart, Hornell: *The Technique of Social Progress.* N. Y., Hen. Holt & Co., 1931. xv and 708 pp., il., $3.60. Esp. chaps. 3, 4, 19, 20, 22. Judicious and lively. See his article hereafter.

160. Hatfield, H. Stafford: *Automaton,* or the future of the Mechanical Man. Lon., Today and Tomorrow ser., 1928. 100 pp. Discusses the econ. problems of inventors, and patent shortcomings.

161. —: *The Inventor and his World.* Lon. 1933 and from Dutton, N. Y., 269 pp., $2.40. Bibliographies, pp. 255-63. A valuable discussion by a physicist and inventor of the present situation of inventors, some fields of inventive progress, and patent law. Reviewed by me in *Am. Jol. of Soc.* 40: 691, 2.

162. Heppner, Ernst: *Juden als Erfinder und Entdecker.* 1913; 125 pp. Biographies.

163. Hobson, John A.: *Evolution of Modern Capitalism.* Rev. ed., 1906; esp. chap. 4, sec. 5, 6.

164. Hulme, E. W.: *Statistical Bibliography in Relation to the Growth of Modern Civilization.* Lon., 1923. Excellent counting of books, journals and patents.

165. Kaempffert, Waldemar: *Invention and Society.* Am. Lib. Asn., Chicago, 1930—Reading with a Purpose, No. 56. 37 pp. On social causation and method of invention.

166. Kuznets, Simon S.: *Secular Movements in Production and Prices.* Houghton Mifflin, Boston, 1930. 536 pp. Pp. 9-35 and 54-8 contain well-supported claims for a tendency toward a retardation of progress in the technology of manufacturing.

167. Mach, Ernst.: *Pop. Sci. Lectures.* Tr. by T. J. McCormack; 3d ed., 1898. Chap. on the part played by accident in invention and discovery.

168. Montmasson, J. M.: *Invention and the Unconscious.* Tr. 1932 by H. S. Hatfield, from *Le Rôle de l'inconscient dans l'invention scientifique,* 1928. Esp. pp. 169-87.

169. Ogburn, Wm. F.: *Social Change,* with respect to culture and original nature. N. Y., Viking Press, 1922. $2. Important on both the causation and results of invention. Cf N 72; and Woodard in the Index.

170. Ravenshear, A. F.: *The Industrial and Commercial Influence of the English Patent System.* 1908. 160 pp. Historical, statistical and intelligent.

171. Rossman, Jos.: *The Psychology of the Inventor*—a study of the patentee. Rev. ed., 1931, x and 252 pp., Inventors Pub. Co., 1266 N. H. Ave., Wash,; $3. Excellent bib. in pp. 237-46. The *best book* that has yet appeared on the sociology of invention, it should be possest by every student of the subject. The author's unique personal acquaintance, with both invention, patents, Sociology and Psychology are reflected in the factual rather than philosophical or historical character of the work, which makes considerable use of important questionnaires. Reviewed by the present writer in *Am. Jol. of Soc.,* 38: 482-4.

172. Sorokin, P.: *Contemporary Sociological Theories.* 1928. Esp. pp. 409-12 on size and density of pop. correlated with inv., and 742-6 on effects.

172.1. —: *Social Mobility.* Eminent inventors live to 74.7 yrs. av. p. 259. Value of mixed cultures, p. 511ff.

173. Stamp, Sir Josiah: *Some Economic Factors in Modern Life.* 1929. Chap. 3, Invention, pp. 89-121.

174. Stern, Bernhard J.: *Social Factors in Medical Progress.* Col. U. Press, 1927. 136 pp. Chiefly and well expounds duplicate discovery and resistance to innovation.

175. Sweeney, H. W.: *Certain Aspects of Inventions in their Relation to Cultural Evolution.* U. of Chicago Master's thesis, unpub., 1929. 266 pp. Hist. compilation on 8 major inventions, and discussions of causation.

176. Tarde, G.: *The Laws of Imitation,* tr. by Elsie Clews Parsons, Holt pub., 1903. An outmoded system of Sociology, on the basis of innovation and imitation. P. 127 quoted.

177. Taussig, F. W.: *Inventors and Money-makers.* Macmillan, 1915. A study of famous inventors of the past.

178. Thompson, Holland: *The Age of Invention.* 1926. Also inc. in *The New Industrialism.*

179. Usher, A. P.: *A History of Mechanical Inventions.* N. Y., McGraw-Hill, 1929, 412 pp., $5. Contains a keen psychological analysis of invention in pp. 1-31, followed by sound and well-illustrated accounts of the history of Mechanics, power sources, timepieces, Leonardo's inventions, printing, textiles and machine tools.

180. Vaughan, Floyd L.: *Economics of our Patent System.* N. Y., Macmillan, 1925. 303 pp. Bib. on pp. 265-77. An excellent work, unique in America. A critical study by an Economist of how patents really work out, with abundant citation of legal decisions. Much on the suppression of patents, and the fortification of monopoly, which is decried.

181. Waffenschmidt, W. G.: *Technik und Wirtschraft.* G. Fischer pub., Jena 1928. 328 pp., il. An evolutionary, philosophical study of the forces of econ. hist., w. bib., by an economist.

182. Winston, Sanford: *Culture and Human Behavior.* N. Y. 1933. Chap. 5: The Accumulation of Culture.

Articles

183. Bernard, L. L.: Interdependence of Factors Basic to the Evolution of Culture; in *Am. J. of Soc.* 31: 1-18.

184. Brinkmann, Carl: Invention. Article in *Encyc. of the Soc. Sciences,* 8: 247-51. Well thot and with an excellent Bib., q. v.

185. Carr, Lowell J.: The Patenting Performance of 1000 Inventors during Ten Years; in *Am. Jol. of Soc.* 37: 569-80, 1932. Showing the high prolificity of the few successful inventors, and the small percentage (0.2%) of the pop. which takes part during a decade.

186. —: Study of 137 Typical Inventors; in *Pubs.* of Am. Sociological Soc. 23: 204-6, 1929. Valuable results of a questionnaire, worked up with careful statistical method.

187. Clerk, Dugald: Discovery and Invention; in *Engineer* 124: 514-5, 1917.

188. De Leener, Georges: Sur la filiation des inventions; in *Bulletin Mensuel,* Instituts Solvay, Archives Sociologiques, 20: No. 322, pp. 653-7, 1912.

189. —: Sur les directions imposées aux inventions techniques par les conditions gén. de la production industrielle; in *do.,* 16: No. 263, 1911.

190. —: Sur les enchainements des inventions techniques et leurs influences sur l'organisation de l'industrie; in *do.,* 14: No. 218.

191. Dickinson, H. W.: The Evolution of Invention; in *Engr.* 157: 107, 8; and w. related articles in *Engg.* 137: 1: 89-91, & N 198.1. On modern inv.

192. Dupréel, E. Sur les conditions de l'invention scientifique; in *Bull.* No. 26 de l'Inst. de Sociologie, 1913, pp. 652-6.

193. Epstein, Ralph C.: Industrial Evolution: Heroic or Systematic?; in *Q. J. of Ec.,* Feb. 1926, pp. 232-72. The latter is the answer.

194. Gilfillan, S. C.: Inventiveness by Nation—A note on statistical treatment; in *Geog. Rev.* 20: 301-4, 1930. Reworks the statistics of Prof. Mark Jefferson, to show the comparative inventive rank of the nations. Reprinted in *Pat. Office Soc. Jol.* 12: 259-67, with additions of correlations by my student, Miss Marian Hurlbut, of the inventive score of the American states

with social conditions. One criticism of international patent statistics, in addition to those given therein, is that patent lawyers might be less scrupulous in some countries, and there encourage patenting an invention all over the world.

195. —: Who Invented It? See N 305.

196. Hart, Hornell: Preliminary Conclusions from a Study of Inventors; in *Pubs.* of Am. Sociological Soc. 21: 191-4, 1927. More detailed than in his book. Reviewed by S. C. G. in *Pat. Off. Soc., J.* 10: 115-9.

197. Hatfield, H. S.: The Encouragement of Invention; in *Psyche,* 10: 90-117, 1930. An excellent British consideration of the inventor-patron relation, and decadence and abuses of the patent system. The burdens of legalism are deplored, but extensions of the patent system are proposed to benefit all introducers, which would much increase patent business and litigation.

198. Henderson, Prof. Sir Jas. B.: Invention as a Link in Scientific and Econ. Progress; in *Sci. Mo.,* Oct. 1927, pp. 347-61. On the methods and social controls of inv. and their possible improvement.

198.1 Instn of Mech. Engrs., *Proc.,* 126: 3-51, 1934. A symposium on inv. by H. W. Dickinson, Hatfield, Gledhill, Ravenshear and numerous commentators.

199. Kaempffert, Waldemar: Invention as a Social Manifestation. Chap. 2, pp. 20-65, of *A Century of Progress,* ed. by C. A. Beard, Harper pub., 1933. An excellent survey of historical and contemporary evidence.

200. —: Our Defective Patent System; in *Outlook* 101: 548-51, 1912.

201. —: Systematic Invention; in *Forum* 70: 2010-18, 1923; and Invention by Wholesale; in the same, pp. 2116-22. An able comparison of contemporary inventing with that of the past century, dismissing the heroic theory of invention, holding that revolutionary work is due to outsiders, and explaining the obstacles to inventions.

202. Kroeber, A. L.: On the Principle of Order in Civilization as Exemplified by Changes of Fashion; in *Am. Anthrop.* 21: 235-63, 1919. Esp. pp. 262, 3.

203. —: The Superorganic; in *Am. Anthrop.* n. s. 19: 163-214, 1917; reprinted w. revision, 1927.

204. Miller, H. S.: Invention, Inevitable or Divine?; in *Pat. Off. Soc. J.* 10: 406-11, 1928. Criticizes my "Who Invented It?", adding my 27th Prin. and a metaphysical Higher Power, and failing to see that I spoke first of the common idea, not the truth of invention.

205. Mitman, Carl W.: Watchmakers and Inventors; in *Sci. Mo.,* July 1927.

206. Ogden, Chas. K.: *Bentham* on Invention; in *Psyche,* Oct. 1929, pp. 102-9. The famous Jeremy B—, successful inventor as well as philosopher, describes the difficulties of inventors around 1787.

207. Ostwald, Wilh.: Systematisches Erfinden; in *Prometheus* 24: 5-8, 17-22.

208. Protection by Patents of Scientific Discoveries. Rept. of the Com'ee on Pats., Copyrights and Trade Marks, Jos. Rossman, Chmn.; Occasional Pub. No. 1, of Am. Asn. for the Advancement of Sci.; N. Y., Science Press, 1934; 45 pp., 50c.

209. Roege, F. W.: Intelligence Measured by Inventiveness; in *Pat. Off. Soc. J.* 10: 206-12, 1928. A good partial correction of N 213.

209.1. Rossman, Jos.: The Correlation of Intelligence and Invention; in *Pat. Off. Soc. J.* 10: 356-9, 1928. Cf N 213.

210 Rossman, J.: Do Engineers Invent, and Why do they Abhor the Label "Inventor"?; in *Technology Rev.*, 24: 117ff., 1931. Results of a questionnaire to engineers in *Who's Who in Engg.*, who had made an invention, 258 replying.

211.—: Engineers as Inventors; in *Pat. Off. Soc. J.* 13: 376-83, 1931.

212.—: The Negro Inventor; in *Pat. Off. Soc. J.* 12: 549-53.

213.—: The Relation of Intelligence to Invention; in *Pat. Off. Soc. J.* 1927, v. 9: Pt. I: The Relation of Invention to the Intelligence of Foreigners, pp. 511-34; and Pt. II, among American states, pp. 565-71. The statistics are usable, but the explanations need the corrections of their author (N 209.1), Roege N 209, and Gilfillan and Jefferson N 194.

214.—: Seasonal Variations in Applying for and Granting Patents; in *Pat. Off. Soc. J.* 11: 99-103, 1929. Other such graphs are in the same *Jol.* of 2: 434 and 3: 394.

215.—: A Study of the Childhood, Education and Age of 710 Inventors. To be published, under possibly changed title, in *Pat. Off. Soc. J.* 1935. Using further data from his questionnaire returned by 710 inventors, the author obtains further information on their boyhood interests, education, and age at first invention, first patent and best invention so far.

216.—: War and Invention; in *Am. J. of Soc.* 36: 625-33, 1931. Patent stat. from 3 countries.

217.— Women Inventors; in *Pat. Off. Soc. J.* 10: 18-30.

219. Severance, B.: A Select Bib. on Inventions and Inventors; in *Libraries* 33: 247-50, 1928.

219.1. Smith, B.: Engineer as Inventor; a stat. study based on *Who's Who in Engg.;* in *Mech. Engg.* 56: 263-5, 1934.

220. Tarde, G.: L'Invention, moteur de l'évolution sociale; in *Rev. Internat. de Sociologie* 10: 562-74, 1902.

221. Tisell, Henrik G.: Undersökning öfver Uppfinnarverksamhetens Variationer. (Inquiry regarding the variations of inventive activity in various fields in Sweden, Germany, France, England, Austria and Hungary, prepared by order of the Royal patent legislation com'ee.) In *Statsvetenskaplig Tidskrift, Extrahäfte,* 1910; 117 pp. A valuable study by a socially minded patent official, which with its numerous tables and graphs may be used by scholars with no knowledge of Swedish. As in the sample we have reprinted, it traces the ups and downs of patenting in very numerous fields and group-inventions, and seeks explanation for the same in industrial history.

222. Winston, Sanford: American Inventors—some bio-social characteristics; in *Social Forces,* 1934 or '5.

223. Woodworth, R. S.: The Mechanism of Progress; in *Sci. Am. Sup.* 69; 279.

224. Wyman, W. I.: Age of Production in Invention and other Fields; in *Pat Off. Soc. J.* 1: 439ff. On inventors' ages.

A BIBLIOGRAPHY ON THE SOCIAL EFFECTS OF INVENTION

The following works listed elsewhere are appropriate to this Bibliography: Douglas N 138, Douglas & Director N 136, Epstein N 193, Haldane N 158. Holland N 270, Montmasson N 168, Ogden N 206, Stamp N 173, Sweeney N 175, Vaughan: *Economics of our Patent System,* N 180, and Waffenschmidt N 181.

Books

226. Baker, Eliz. F.: *Displacement of Men by Machines.* Col. U. Press 1933; 244 pp. Case of the printing trade.

227. Barnes, H. E.: *Living in the 20th Century.* 1928.

228. Chase, Stuart: *Men and Machines.* Macmillan, N. Y., 1929; 354 pp. Good on the human effects of invention, esp. on workers. Bib.

229. Dalton, Hugh: *Some Aspects of the Inequality of Incomes in Modern Communities.* London 1920; pp. 210-17. By an economist.

230. Diesel, Eugen: *Völkerschicksal u. Technik.* Cotta pub., Stuttgart u. Berlin, 1930. 138 pp.

231. Douglas, Paul H.: *Theory of Wages.* Macmillan pub., 1934; 659 pp., $5. For the effect of invention on wages. Pp. 209-16, using our data.

232. Hausleiter, Leo: *The Machine Unchanged.* Appleton-Century, 1934, 376 pp. By a German engineer-economist.

233. Launhardt, Wilh.: *Am sausenden Webstuhl der Zeit,* Uebersicht über die Wirkungen der Entwicklung der Naturwissenschaften u. der Technik auf das gesamte Kulturleben. 3d ed., 1910. By a thotful engineer.

234. Mumford, Lewis: *Technics and Civilization.* Harcourt Brace pub., 1934. 507 pp., il. Historical evidence. List of inventions from the 900s. Excellent bib.

235. Noiré, L.: *das Werkzeug u. seine Bedeutung für die Entwicklungsgesch. der Menschheit.* 1880. Philos. & tech.

236. Ogburn, Wm. F.: *Social Change.* See N 169.

237.—: with the assistance of S. C. Gilfillan: The Influence of Invention and Discovery; Chap. 3, 45 pp. of v. 1, *Recent Social Trends in the U. S.,* Report of the President's Research Committee on Soc. Trends; 2 vols., 1568 pp., McGraw-Hill pub., N. Y. 1933.

In the same vol. Prof. Ogburn with the assistance of Clark Tibbitts discusses the influence of inventions on the family; esp. pp. 664-72. Cf. N 240.

237.1.—: *Living with Machines.* 16 pp., Am. Lib. Asn., 1933. W. bib.

237.2.—: *You and Machines.* 58 pp.; Nat'l Capital Press, 1934. Written for CCC camps.

238. Pigou, A. C.: *Wealth and Welfare.* Lon., 1912; pp. 716-24. Compares labor-saving and capital-saving inventions, for their effects on wages.

239. Tugwell, R. G.: *Industry's Coming of Age.* Col. U. Press, 1927; 283 pp.

240. Willey, M. M., & Rice, S. A.: The Agencies of Communication; Chap. 4 in *Recent Social Trends,* v. 1. (N 237). Soc. and econ. effects of invs.

Articles

242. Boas, Geo.: In Defense of Machines; in *Harper's,* Je. '32, pp. 93ff. Effects on work and recreation.

243. Brown, Percy S., ed.: The Second Industrial Revolution and its Significance; in *Annals Amer. Acad.,* May 1930, a symposium.

244. Clemen, R. A.: Effect of Scientific Inventions upon Economic Trends in *Facts and Factors in Econ. Hist.,* 1932, pp. 666-91.

245. Gilfillan, S. C.: Inventions and Discoveries during 1931; in *Am. Jol. of Soc.* 37: 868-75, 1932. Similarly,

246.—: Inventions and Discoveries; in *do.* 38: 835-44, 1933. Classified lists with general considerations. These articles, together with earlier lists in the same series, by W. F. Ogburn and Clark Tibbitts, were pub. in the May numbers of the *Am. Jol. of Soc.,* and also inc. in separate annual volumes entitled *Social Changes in* 19—. The general considerations from the second article, on the repartition of inventions, were reprinted in the *Pat. Off. Soc. J.* 15: 567-9.

247. Greenberg, A. S.: Graphic Study of the Econ. Relations of the Pat. System; in *Pat. Off. Soc. J.,* Oct. '26, pp. 64-8. For its stat., not reasoning.

248. Invention and the Demand for Labor; in *Pat. Off. Soc. J.* 15: 653-7, from Je. 1933 letter of the Nat'l City Bk. of N. Y.

249. Jome, H. L.: Effect of Inventions on Industry; in *Cur. Hist.* 30: 585-92.

250. Thomas, Jas. S.: What the Machine is Doing to Mankind; in *Franklin Inst. J.,* Mch. 1933, pp. 247-72.

A FEW WORKS ON THE PSYCHOLOGY OF INVENTION

To the following should be added Hart N 159, Hatfield N 160, Le Chatelier N 273, Montmasson N 168, Roege N 209, Rossman N 171, Taussig N 177, Usher N 179, and Woodworth N 223.

252. Cleeton, G. U.: Originality, a summary of experimental literature; *in Jol. of Abnormal & Soc. Psych.* 21: 304-315, 1926.

253. Ostwald, W.: *Erfinder u. Entdecker.* 99 pp.

254.—: *Grosse Manner, Lpz.* 1909; 424 pp. Biographies and conclusions. Tr. by Dufour as *Les Grands hommes.*

255. Ribot, Th. A.: *Essai sur l'imagination créatice.* 1900.

256. Tanner, Amy E.: Certain Social Aspects of Invention; in *Am. Jol. of Psych.,* 26: 388-416, 1915.

257. With, Emile: *Les Inventeurs et leurs inventions.* 1864; 261 pp. A dictionary of biography with much information on the then status of inventors.

BIBLIOGRAPHY OF INVENTIVE METHOD

A short Bibliography on Primitive Invention will be found in our companion volume.

The following works listed elsewhere are appropriate to this Bibliography: Grosvenor, N53; Hatfield N160 (esp. pp. 79-89), and N161; Protection . . . . . of Scientific Discoveries N208; Kaempffert N201; and Olken N139.

259. *Coöperative Industrial Research.* A booklet by the Dept. of Mfg., Chamber of Commerce of U. S., Wash., 1925.

260.—: (ditto) An Acct. of the Work of Research Asns. under the govt. scheme, w. a preface by Earl Balfour. Lon., H. M. Sta. Office, 1927. 46 pp.

261. Dickinson, Z. C.: *Industrial and Commercial Research;* Functions, finances, organization. Mich. Bus. Studies, Nov. 1928. 70 pp. 50c.

262.—: *Suggestions from Employés.* Do., Aug. 1927. 60 pp.

263.—: *Research Activities of Business Organizations.* Gen. Mgmt. Ser., No. 81, of Am. Management Asn. 35pp., 1929, $1.

263.1. —: Suggestion Systems; in *Mech. Engg.* 56:671-5, 1934. Do. in 1932, 3, by P. L. Stanchfield & D—; in *Personnel Jol.* 13:197-203.

264.—: Suggestions from Workers: schemes and problems; in *Q. J. of Econ.*, 46: 617-43, 1932.

264.1. Douglas, Jas.: *Untechnical Addresses on Tech. Subjects.* 2d ed., NY. 1908. Esp. the chaps. on the Characteristics and Conditions of Tech. Progress in the 19th Cen., pp. 1-38; and Secrecy in the Arts, 127-64.

265. *Erfindungswesen in der USSR, das;* pub. in Russia. See esp. Organizations for the Realization of Inventions and their Work; in 1: No. 10: 32-6, 1931.

266. Fleming, A.P.M., and Pearce: *Research in Industry the Basis of economic progress.* Lon., 1922. Bib. on pp. 221-36.

267. Gammeter, J. R.: We Make Inventing an Every-day Job; in *System,* Feb. '25, p. 181ff. Methods of the Goodrich tire laboratory.

268. Holland, Maurice, with Pringle, H. F.: *Industrial Explorers.* Harper pub., 1928. 359 pp.

269. Holland, Maurice: Research in Europe—a comparative study of the national and industrial organization. Mimeographed paper presented Nov. 21, 1924, to Div. of Eng. & Indus. Research, Nat'l Research Council.

270.—: Research, Science and Invention; chap. 16, pp. 312-34, of *A Century of Progress,* ed. by F. W. Wile. Doubleday pub., 1928.

271. Hopkins, N. M.: *Outlook for Research and Inventions.* 1919. Covers the war and later situation.

272. Kaempffert, W. et al.: *Progress*—Amer. Institute of the City of N. Y. Exposition of Inventions . . . . 1925. Invention Today, a Symposium. Good articles, upholding corporative inventing.

273. LeChatelier, H. L.: *Science et Industrie.* Flammarion pub., 1925. By an eminent inventor.

274. *Popular Research Narratives;* pub. by Engg. Foundation, N. Y., 1924ff. Collections of interesting stories by men concerned, of how invention proceeds today.

275. Warner, E. P.: The Nature of Invention; in *Harv. Grads. Mag.* 31: 310-17, 1923. Tells the importance of knowledge and work and the futility of the outsider.

A SELECT BIBLIOGRAPHY OF HISTORIES OF INVENTION

The following works cited elsewhere, are also appropriate to this Bibliography: Fiske ftN *3*, Lefebvre des Noëttes N 49, Usher N 179, and With N 257.

Also the journals *Trans.* of the Newcomen Soc. for the Study of the History of Engineering and Technology (biennial, 20 s., Science Mus., S. Kensington, London S. W. 7, or 29 W. 39th st., N. Y.), *Isis* (qly., Hist. of Sci. Soc., F. E. Brasch, Cor. Sec., Lib. of Cong., Wash.), *Beiträge zur Geschichte der Technik u. Industrie,* and *Geschichtsblätter für Technik, Industrie u. Gewerbe.*

277. Darmstaedter, L: *Handbuch zur Geschichte der Naturwissenschaften u. der Technik,* in chron. Darstellung. 2nd ed., 1908, 1262 pp.

278. Dudeny, H. E.: Antiquity of Modern Inventions; in *Strand* 45: 388-92.

278.1. Espinas, A.: *Les Origines de la technologie;* étude sociologique, 1897.

279. Feldhaus, F. M.: der Ingenieur im Altertum, im Mittelalter u. in der neueren Zeit; in *Der Ingenieur,* Stuttgart, 1910, pp. 7-19. Cf. also pp. 36-41.

280.—: der Laie als Erfinder; in *Geschichtsblätter für Technik,* 5: 6-53.

280.1.—: *Ruhmesblätter der Technik.* 2d ed., Lpz., 1926, 2 v., il.

280.2.—: *Technik der Vorzeit.* Lpz. 1914, 1400 pp., 873 ils.

281.— and Corey, Lewis: Machines and Tools, article in *Encyc. of the Soc. Scis.* How transformation of factory methods was brot about by successive needs.

281.1. Fournier, Ed.: *Le Vieux-neuf;* histoire ancienne des découvertes modernes. 1859. Well documented; a little too devoted to showing how inventions have been antedated.

282. Hammond, Jn. L. LeB. & Barbara: *The Rise of Modern Industry.* London, 1925. Historical.

283. Iles, Geo.: *Inventors at Work,* with chaps. on discovery. 1906, Doubleday Page.

284. Jobard, J.B.A.M.: *Les Nouvelles Inventions aux Expositions univ.* 2 v., Brussels, 1857.

285. Kaempffert, Waldemar, ed. and largely author: *A Popular History of American Invention.* 2 v., large 8o, 500 il., Scribner, 1924, $10. Shortened in his *Modern Wonder Workers.* One of the best of the many histories of invention. Covers some European as well as American developments of the last century or two.

286. Neuburger, Alb.: *Erfinder und Erfindungen.* Berlin, Ullstein pub., 1913. 275 pp. Covers the last century, and upholds evolutionism.

287.—: *The Technical Arts and Sciences of the Ancients.* Tr., Macmillan, 1930; 518 pp., 676 il.

287.1. Roe, Jos. W.: *English and American Tool Builders.* McGraw-Hill pub. 1916.

288. *Science and Invention.* v.8 of *The Making of Amer.*, R. M. LaFollette, ed., Chicago, 1900-06. Various descrip. and hist. articles.

289. Zimmer, G. F.: *Engineering of Antiquity.* Probsthain pub., 1913.

SCIENCE BIBLIOGRAPHY OF THE AUTHOR

291. The Coldward Course of Progress; in *Pol. Sci. Q.* 35:393-410, 1920; reprinted with amendments in *Hist. Outlook* 12: 8-15. Accepted by Ellsworth Huntington (N *43*) and in his Temperature and the Fate of Nations; in *Harper's Mag.* 157: 361-8 and in his (and Williams' & Van Valkenburg's) *Economic and Social Geography,* Wily pub. 1933, pp. 141, 2; and by Camille Vallaux, Franklin Thomas, Chas. Hodges, V. Stefansson, Wallis & Willey. (N 305) etc. To be reprinted by Kimball Young in his forthcoming *Source Book for Sociology.*

292. Early Steamboats of the Chesapeake; in *Balto. Sun,* May 17, 1931, Mag. sec., pp. 10-11.

293. European Political Boundaries; in *Pol. Sci. Q.* 39: 458-84, 1924; reprinted revised in *Hist. Outlook* 16: 65-75 and accepted by E. Huntington, and Chas. Hodges: *Background of International Relations,* 1931 (chart reproduced).

294. The First Seagoing Auxiliary (The *Savannah*); in *Yachting* 48: 54-7, 114, 5; July, 1930.

295. The Future Home Theater; in *Indep.* 73: 886-91, 1912.

295.5. Housekeeping in the Future; in *Indep.* 72: 1060-2, 1912.

296. *Inventing the Ship.* A study of the inventions in her history between floating log and rotorship. When, where, why, by whom and how were they made? Illustrating the principles discust in the present volume. With 80 illustrations, Bibliographies, Notes and Index. A self-contained volume, intended for all who care for the history of the ship, whether or not concerned with the social aspects of invention.

297. Invention in the History of the Ship. Paper summarizing this book. read before a section of the Am. Sociological Soc. and pub. in their *Publications,* 23: 200-3, 1929; reprinted in *Pat. Office Soc. J.*

298. Invention in the History of the Ship. Our companion volume, without corrections, additions, Notes and bibliographies, was pub. serially in the *Marine News,* Aug. 1928—March, '29.

299. Inventions and Discoveries (during 1931 and '32); in *Am. Jol. of Sociology* 37: 868-75, 1932, and 38: 835-44, 1933; reissued in volumes entitled *Social Changes* (of 1931 and '32). Classified and discust lists of the most socially portentous inventions and discoveries, preceded by general discussions of the prediction of social effects, of which the second was reprinted in the *Pat. Off. Soc. J.* (N 246).

300. Inventiveness by Nation; reprinted augmented as Inventiveness by Nation and State. See N 194.

301. (Museum) Has Model of Locomotive Invented in 1795 (Fitch's); in *Detroit News,* May 18, 1930, feature sec., p. 3.

302. A Railroad Man's Museum. *Ill. Cen. Mag.,* 1929.

303. The Size of Future Liners, in *Indep.* 74: 541-3, 1913.

304. Social Principles of Invention. The present work, but lacking most of the notes, and the bibliographies, graphs, index and numerous passages, was published serially in the *Patent Office Soc. Jol.,* beginning Jan., 1934.

305. Who Invented It? in *Sci. Mo.,* 25: 529-34, 1927. Reprinted in *Jol. of the Pat. Office. Soc.* 10: 215-25, *Reader's Digest,* and W. D. Wallis and Willey's *Readings in Sociology,* pp. 65-72. (Knopf pub., 1929).

306. Signed Book Reviews of Bowman's *New World,* Hart's article: Preliminary Conclusions from a Study of Inventors (*Jol. of Pat. Off. Soc.,* 10: 115-9), Hansen's *Cycles of Prosperity and Depression,* Hatfield's *The Inventor and his World* (N 161), Huntington's *Character of Races, West of the Pacific* and *Builders of America;* Rea's *The Museum and the Community;* Hawks' *The Romance of Transport;* Rossman (N 171), Bouthoul's *L'Invention,* Stefansson's *Northward Course of Empire,* and Stockton's *International Moulders'* Union.

These reviews were published in: *Am. Jol. of Sociology, Pol. Sci. Q., Hist. Outlook, Sewanee Rev., & Pat. Off. Soc. Jol.*

INDEX

NUMBER CODE AND ABBREVIATIONS: Simple numbers indicate pages. All numbers after an N refer to Citational Notes, forming the first Appendix of this volume, pp. 265-74. Numbers in *italics,* preceded by ftN, as ftN *140,* refer to a Footnote, which can be readily located at the foot of one of the text pages. (Both kinds of notes form a single numerical series.) Often the *most important reference* is distinguished by an asterisk, as (page) *17, or ftN **56*. 50, 1, 4-6 would mean pages 50, 51 and 54 to 56.

Prin. refers to one of the Social Principles of Invention listed in pp. 5-13 of Chap. I, and indexed hereafter under Principle. Names of ships and of books and periodicals, the latter two often abbreviated, are in *italics*.

PRINTED IN THE U. S. A.

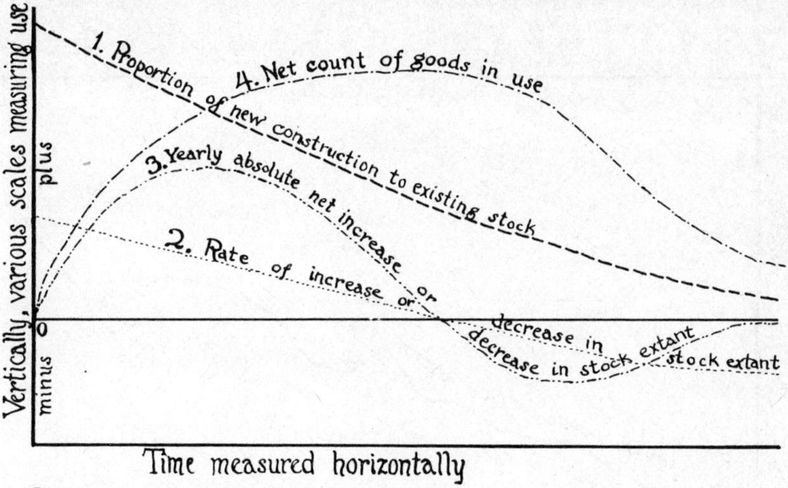

Graph 1: THE LIFE CYCLE OF AN INVENTION
(Hypothetical Typical)

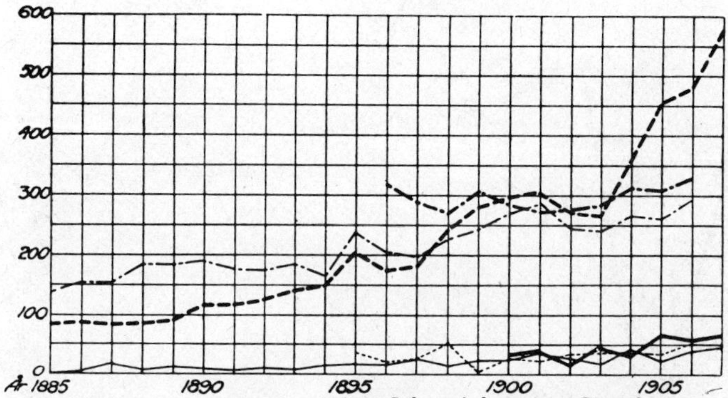

Applications, Sweden
Patents, "
Applications, Germany
" , England
Patents, France
" , Austria

Graph 2: SHIPBUILDING PATENTS OR APPLICATIONS, IN 4 FOREIGN COUNTRIES, 1885-1907. From Tisell (N 221)

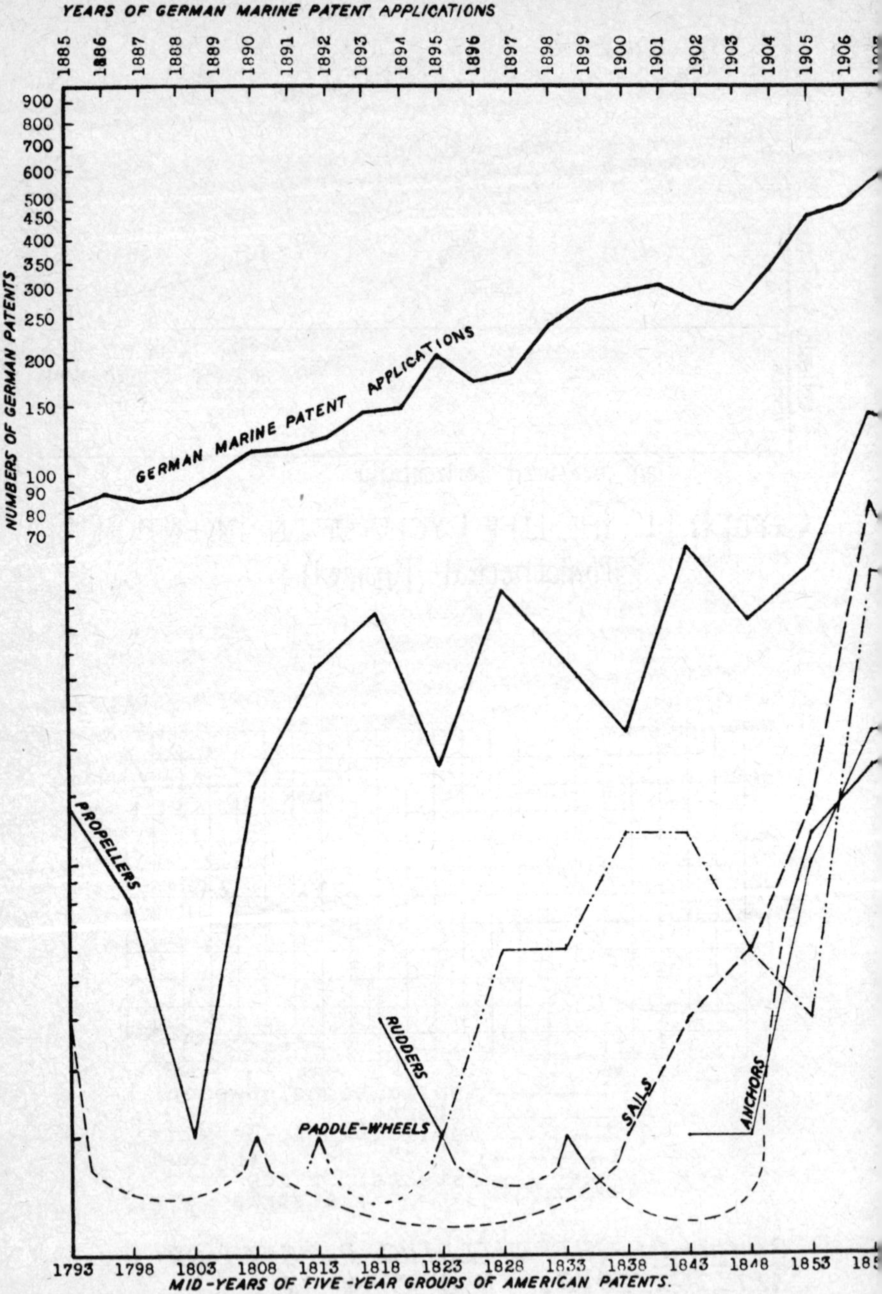

Graph 3. American Patents on Propellers, Paddle-Wheels, Rudders, Sails and Anchors, 1790 - 1928,

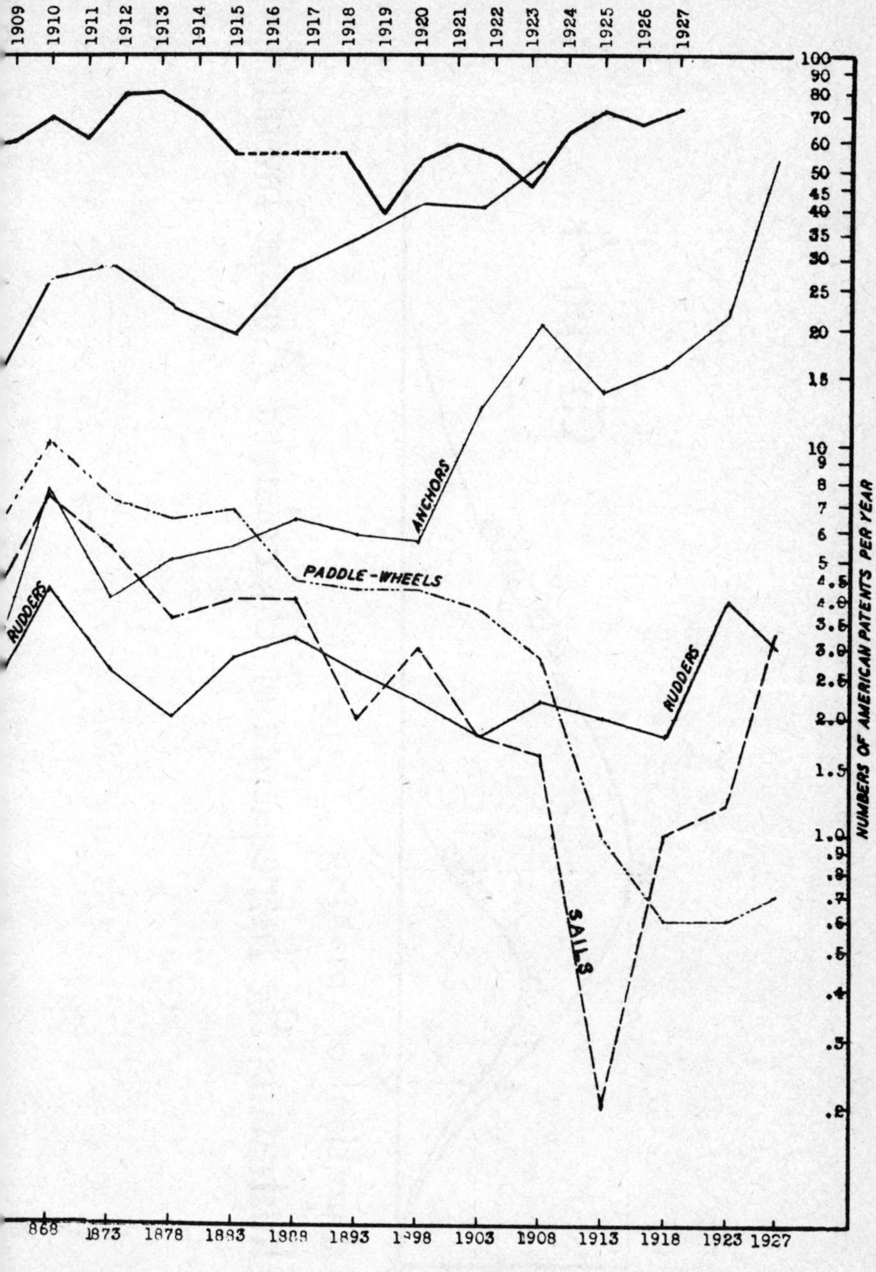

and German Marine Patent Applications, 1885 - 1927.
Patent numbers in logarithmic scale. See Ft.N. 14.

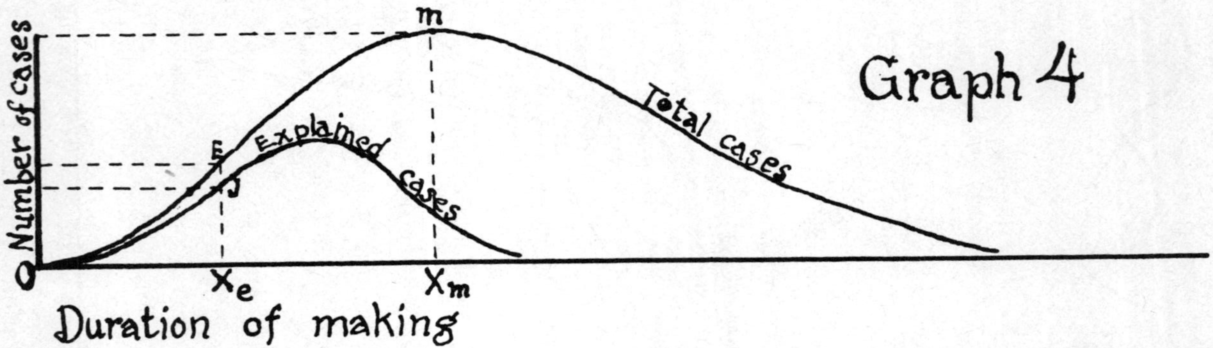

Graph 4

Illustrating the Infrequency of Unexplained Abrupt Inventions

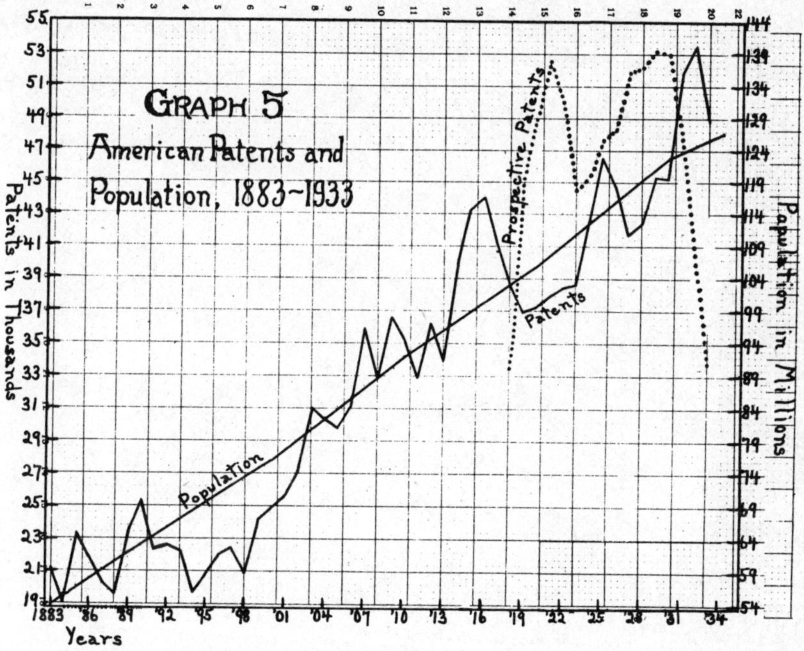

Showing the failure of patenting to increase per capita, despite the presumptive growth of inventiveness. The dotted line labeled Prospective Patents measures patent applications in the same years, reduced to 56.2%, which is the proportion of patents granted, to applications received during those years. (The patents granted in those years had been supplemented by 56% of the increase (42,250) of applications on hand awaiting action.) This dotted line is added as an alternative measure of patenting, because it is more promptly responsive to business conditions, and because the flood of applications during the prosperous 1920s swamped and slowed up the granting of patents. The Patents curve has a variable lag after that applications curve, due to Office and applicants' delays. Previous to about 1885 patents were increasing faster than population. Calendar years are shown after 1900, from <u>Stat. Abstract</u>; fiscal years before, from Commissioners' Reports. See our text, pp. 109 ff.